THE POETS OF
TIN PAN ALLEY

THE POETS OF TIN PAN ALLEY

A History of America's Great Lyricists

PHILIP FURIA

New York Oxford · Oxford University Press 1990

Oxford University Press

Oxford   New York   Toronto
Delhi   Bombay   Calcutta   Madras   Karachi
Petaling Jaya   Singapore   Hong Kong   Tokyo
Nairobi   Dar es Salaam   Cape Town
Melbourne   Auckland

and associated companies in
Berlin   Ibadan

Copyright © 1990 by Philip Furia

Published by Oxford University Press, Inc.,
200 Madison Avenue, New York, New York 10016

Oxford is a registered trademark of Oxford University Press

Library of Congress Cataloging-in-Publication Data
Furia, Philip, 1943–
The poets of Tin Pan Alley : a history of America's great lyricists / Philip Furia.
p.   cm.   Includes bibliographical references and index.
ISBN 0-19-506408-9
1. American poetry—20th century—History and criticism.
2. Songs, English—United States—History and criticism.
3. Popular literature—United States—History and criticism.
4. United States—Popular culture—History—20th century.
5. Popular music—United States—History and criticism.
6. Lyric poetry—History and criticism.
7. Lyricists—United States.   I. Title.
PS309.L8F8   1990   782.42164′026′8—dc20   90-35937

9 8 7 6 5 4 3 2

Printed in the United States of America
on acid-free paper

For Karen,
It had to be you

Preface

This book began while I was a Fulbright professor at the University of Graz in Austria. When my students in a course on modern American literature, art, and music asked me about popular songs during the 1920s and '30s, I first professed to know nothing about the subject. Then I realized that the songs I had loved all my life, the standards of Rodgers and Hart, the Gershwins, Berlin, and Porter, had once been the popular songs of their time. It was when I tried to satisfy the curiosity of my Austrian students by typing up the lyrics of some of those songs that I realized how truly poetic they were. My first thanks therefore go to my students and colleagues at the University of Graz for sparking and then sharing my enthusiasm for this subject in its earliest stages. I must also thank my students at the University of Minnesota, particularly the diverse group of students I have had the opportunity to teach through Continuing Education and Extension, for helping me develop my ideas about American popular songs.

My colleagues, Marty Roth and Michael Hancher, gave generously of their time to read my entire manuscript with great care, never allowing their enthusiasm for my subject to relax their demands for clarity of insight and coherence of argument. To the extent the final version of the book meets those demands, it is indebted to them. Thanks, too, to other colleagues who read portions of the manuscript at crucial stages of its development—Kent Bales, Edward Griffin, Peter Reed, and George T. Wright. Since this is a book I have tried to write for the large audience that loves American popular song, I am grateful too for the careful readings the entire manuscript received from Tony Hill, from Bob Lundegaard, and from Steve Davis, who, along with Steve Benson, invited me to discuss American popular song in several of their programs on KUOM Radio.

I also wish to thank Sheldon Meyer of Oxford University Press for his editorial suggestions, Stephanie Sakson-Ford for her careful editing of the manuscript, and Kathy Erickson of the Lazear Literary Agency for her help in seeing the book through to publication. Thanks, too, to Laurie Patterson, who guided me through the labyrinth of word processing. For their guidance through the maze of copyright permissions, I would like to thank Fred Ahlert, Jr., Lewis Bachman, David Bogart, Jeffrey Brabec, Lynnae Crawford, Jeanne Fong, Sidney Herman, Theodore Jackson, Eula Johnson, Donald Kahn, Florence Leeds, Averill Pasarow, Lourdes Richter, and Jack Rosner and his staff at Warner Brothers Music. Support for my research was provided by the Graduate School of the University of Minnesota and the College of Liberal Arts, which enabled me to do research at the Library of Congress and the New York Public Library's Performing Arts Division at Lincoln Center.

From beginning to end, this book owes numerous debts to Les Block, academic colleague, consummate jazz pianist, and local impresario who has included me in his efforts to keep these great songs alive and kicking. I owe an old debt to my Uncle Eugene, who for years heroically taught me music, and to my mother and father, who endured my enthusiasm for rock and roll but also introduced me to *their* music, which by then had become the standards of American popular song. I hope that I have done the same for my sons, Peter and Nick, who, while still preferring *their* music, have listened to mine with good humor and occasional enthusiasm during the writing of this book. In the dedication I have acknowledged the help of my wife, who listened to the songs with me, discussed the lyrics, read and reread chapters and revisions of chapters, and, most of all, radiated the wit, grace, and urbanity that, I came to learn, is what all those songs are about.

Minneapolis   P. F.
February 1990

Contents

1. Blah, Blah, Blah, Blah Love: Alley Standards, *3*

2. After the Ball: Early Alley, *19*

3. Ragged Meter Man: Irving Berlin, *46*

4. Ragged and Funny: Lyricists of the 1920s, *72*

5. Funny Valentine: Lorenz Hart, *95*

6. 'S Wonderful: Ira Gershwin, *126*

7. The Tinpantithesis of Poetry: Cole Porter, *153*

8. Conventional Dithers: Oscar Hammerstein, *181*

9. Paper Moons: Howard Dietz and Yip Harburg, *195*

10. Fine Romances: Dorothy Fields and Leo Robin, *213*

11. Hip, Hooray, and Ballyhoo: Hollywood Lyricists, *231*

12. Swingy Harlem Tunes: Jazz Lyricists, *244*

13. Midnight Sun: Johnny Mercer, *263*

Notes, *283*
Acknowledgments, *291*
Index, *309*

THE POETS OF
TIN PAN ALLEY

The popular song of the past half-century had the largest impact on American culture of any so-called art form. Why, for God's sake, the popular song IS American culture!

IRVING CAESAR

1

Blah, Blah, Blah, Blah
Love: Alley Standards

A song without words is just a piece of music.

JULE STYNE

Mrs. Oscar Hammerstein, so the story goes, once overheard someone praise "Ol' Man River" as a "great Kern song." "I beg your pardon," she said, "But Jerome Kern did not write 'Ol' Man River.' Mr. Kern wrote *dum dum dum da;* my husband wrote *ol' man river.*" It's easy to understand her frustration. While the years between World Wars I and II have long been hailed as the "golden age" of American popular song, it is the composers, not the lyricists, who always get top billing. "I love a Gershwin tune" too often means just that—the tune—even though George Gershwin wrote many unlovable tunes before he began working with his brother Ira in 1924. Few people realize that their favorite "Arlen" songs each had a different lyricist—Ted Koehler for "Stormy Weather," Yip Harburg for "Over the Rainbow," Johnny Mercer for "That Old Black Magic." Only Broadway or Hollywood buffs know which "Kern" songs get their wry touch from Dorothy Fields, who would flippantly rhyme "fellow" with "Jello," and which of Kern's sonorous melodies got even lusher from Otto Harbach, who preferred solemn rhymes like "truth" and "forsooth." Jazz critics sometimes pride themselves on ignoring the lyrics to Waller and Ellington "instrumentals," blithely consigning Andy Razaf or Don George

3

to oblivion. Such Alley wordsmiths are forgotten even when their lyrics made all the difference: the most famous of all popular songs lay in limbo for years as a piano rag until Mitchell Parish put words to it, yet it is Hoagy Carmichael we think of when we hear "Star Dust."

The answer to the perennial question, "Which comes first—the music or the words?" seems all too clear. "In a well-wrought song," Susanne Langer has said, "the text is swallowed, hide and hair." Even in art song, where an existing poem is set to music, "the poem as a work of art is broken up. Its words, sound and sense alike, its phrases, its images, all become musical material."[1] Thus even when the words are actually composed first, they still seem of only secondary importance.

If music swallows words in art song, how much more ravenous it must be in popular song, where words are written to already composed music. To Anthony Burgess the consumption seems so thorough he can argue that in a song lyric "what is said is not of great importance." While Burgess pays lip service to the notion that in a good song words and music must "affirm a true marriage of equal partners," the sort of marriage he has in mind seems to be the old-fashioned kind with the lyricist, like a dutiful wife, "graciously obscuring the light of the words" to bring out and clarify the pattern of the music.

Because he regards the lyricist's role as playing second fiddle to the composer, Burgess, like many people, draws a firm distinction between song lyrics and poetry. "Poetry demands the concentration of the reader or listener on content, on originality of imagery or verbal trope; the true lyric deliberately damps the striking image." Instead of providing the "verbal shocks" of poetry, the lyricist's art, for Burgess, consists of the "matching of long vowels or diphthongs to long notes, the disposition of primary and secondary syllabic stress, and the management of climax."[2]

However valid that distinction might be for poetry and song in general, it leaves Ira Gershwin, Cole Porter, and other lyricists of the golden age in limbo, since their lyrics, far from dampening witty tropes and striking imagery, bristle with the "verbal shocks" of poetry. Even Burgess confesses to finding Lorenz Hart's lyrics "brilliant" and quotes such gems as

> Beans could get no keener re-
> ception in a beanery.
> Bless our mountain greenery home!

While such a lyric skillfully matches long vowels to long notes and deftly disposes verbal to musical stresses, it also demands, like poetry, that the listener concentrate on the witty image and what Burgess himself calls the "ingenious" rhyme.

Yet such lyricists themselves have resisted attempts to celebrate these poetic qualities in their songs. George Balanchine, who choreographed several Rodgers and Hart shows in the 1930s, called Lorenz Hart the "Shelley of America" and urged him to publish his lyrics, without music, as a book of poetry. Like most lyricists of his era, however, Hart never seems to have considered such a project, regarding his art as one that was utterly dependent upon the music it set. Usually he could not even begin working until Rodgers had completed the melody, and, in typical Tin Pan Alley fashion, often started with a "dummy" lyric—banal, nonsensical, or even salacious phrases that helped him remember the rhythm and contour of the melodic line. Hart, in fact, acknowledged his subordinate role by breaking with theatrical tradition to allow the composer's name to precede his. Thus we speak of a Gilbert and Sullivan operetta but a Rodgers and Hart musical.

Even the few lyricists who did publish their lyrics separately made sure no one would accuse them of impersonating a poet. Ira Gershwin entitled his collection, with mock-pomposity, *Lyrics on Several Occasions by Ira Gershwin, Gent.* and posted a disclaimer as foreboding as Mark Twain's "Warning" to readers looking for a motive, moral, or plot in *Huckleberry Finn:*

> Since most of the lyrics in this lodgment were arrived at by fitting words mosaically to music already composed, any resemblance to actual poetry, living or dead, is highly improbable.[3]

Even granting that much of the lyricist's craft rests in the artful fit of word to music, such an absolute distinction obscures the fact that lyrics also employ the elements of poetry—rhyme, imagery, metaphor—and some lyrics use these elements intricately enough to merit the same attention we give to poetry. A glance at any anthology will reveal that some of the most famous "poems" of the English language, such as "Drink to Me Only With Thine Eyes" and "A Red, Red Rose," are song lyrics—not "art" songs but lyrics, like Ira Gershwin's, that were set to already composed music.

There is simply no simple distinction between lyrics and poetry. Some lyrics, such as Stephen Foster's, so efface themselves before music that we would never try to "read" them as poetry. Others, like those of Robert Burns, present such subtle poetic features that we sometimes forget we are reading song lyrics. Occasionally the resemblance of lyrics to poetry, far from being "highly improbable," is so close it is hard to tell them apart. E. Y. "Yip" Harburg remembered how one day in high school he and a classmate found they shared an enthusiasm for poetry, especially for society verse. When Harburg recited some of his favorite poems, W. S. Gilbert's "Bab Ballads," the classmate informed him that those "poems" were actually song lyrics. "There's music to it?" asked an incredulous Harburg. "Sure is," replied the classmate, whose name was Ira, and invited him over to the Gershwin home to listen to Gilbert and Sullivan records on the family Victrola. "There were all the lines I knew by heart, put to music!" Harburg recalled, "I was dumbfounded, staggered."[4]

One of their favorite books, he noted, was Carolyn Wells' *Vers de Société Anthology*. It contains hundreds of witty poems, from the elegant Cavalier works of Robert Herrick and Richard Lovelace to nineteenth-century verse by Ernest Dowson and Lewis Carroll. Yet many of those "poems" were song lyrics, and one frequently comes across a line that could pass for an ancestor of a Gershwin lyric, from "Your tiny little nose that turns up so pert and funny" to "I wonder if you wonder if I wonder if you wonder." The resemblance is hardly accidental, since society verse is based upon the same principles that underlie the lyrics of

Hart, Gershwin, and Porter—principles spelled out by Carolyn Wells in the introduction to her anthology.

The "great distinction" of *vers de société*, she explains, is "ease" and "playful spontaneity." While it treats its subject with sophistication, the language is never formal and elevated but "terse and idiomatic, and rather in the conversational key." The rhymes are "frequent" and the rhythm "crisp and sparkling." Society verse, she cautions, must never be "ponderous" and sentimentality must be avoided: "enthusiasms are modified, emotions restrained." Its tone should be "playfully malicious," "tenderly ironical," or "satirically facetious."[5]

But what better exemplifies those principles than a lyric by Lorenz Hart?

> When love congeals, it soon reveals
> the faint aroma of performing seals,
> the double-crossing of a pair of heels,
> I wish I were in love again!

Just as such a lyric closely resembles society verse, it is fundamentally different from the traditional song lyric, as described by Burgess, that "damps the striking image" and avoids the "verbal shocks" of poetry.

Yip Harburg and Ira Gershwin were not alone in their youthful admiration of society verse. Along with Howard Dietz, Dorothy Fields, and many other lyricists, they began their careers as writers of "smarty verse," their highest aspiration to place a poem in magazines like *The Smart Set*, *Vanity Fair*, or that pinnacle of verbal wit, Franklin Pierce Adams' (F. P. A.'s) column, the "Conning Tower," in the *New York World*. When these aspiring poets turned to the more lucrative art of songwriting, their lyrics still were rooted in *vers de société*. One of Ira Gershwin's wittiest lyrics, "Tschaikowsky," from his collaboration with Kurt Weill on *Lady in the Dark* in 1941, was based on a poem, consisting of the names of fifty-one Russian composers, which the fledgling poet had placed in *Life* magazine back in 1924. Gershwin was also inordinately pleased to find that some of his "improbably" poetic lyrics were selected for inclusion in the *Oxford Anthology of Light Verse,* along with lyrics by Cole Porter, Howard Dietz, Yip Harburg, and others.

The golden age of popular song, as John Updike reminds us in his tribute to Cole Porter, was also a "heyday of light verse: there were book reviews in verse, and sports stories; there were droll ballades and rondeaux and triolets. The plenitudinous newspapers and magazines published Don Marquis, F. P. A., Louis Untermeyer, Arthur Guiterman, Christopher Morley, Dorothy Parker, Ogden Nash, E. B. White, Morris Bishop, and Phyllis McGinley."[6] In some of those poems, the distinction between verse and song is virtually erased:

> If I were only dafter,
>   I might be making hymns,
> To the liquor of your laughter
>   And the lacquer of your limbs.

How different is Witter Bynner's deft poetic daftness from Leo Robin's irreverent hymn to limbs?

> Venus de Milo was noted for her charms,
> but strictly between us,
> you're cuter than Venus
> and what's more you got arms!

Here the line between verse and lyric is a purely metrical one: Bynner's poem adheres to a regular metrical pattern, while Robin fits his meter to the irregular pattern of Ralph Rainger's musical accents. Such difficult fitting makes the witty achievements of the lyricist all the more admirable. "A light-verse writer," as Updike points out, "is not constrained to extend his inspiration through enough refrains to exhaust the chorus, to shape his syllables toward easy vocalization." Thus when Franklin Pierce Adams, the verbal wizard of the "Conning Tower," tried to turn his society verse talents to lyric writing, he found "This method of the lyrist" (fitting words to music) "infinitely harder."[7] Yet following a musical rhythm with a metrically uneven line can have the advantage of producing, as Robin's does, a lyric with more colloquial ease than poetry—right down to the vernacular punch of "and what's more you got arms!" That conversational phrasing, in turn, makes the clever rhymes and imagery even more surprising when they seem to emerge from everyday American speech.

By blending the rigors of light verse with those of lyric writing, the songs of the golden age sparkled with a poetic wit that few songs, before or since, have displayed. Not all the lyrics of the age glitter, of course. Take, for example, a lyricist such as Oscar Hammerstein, who even at the height of the age was writing such sonorous paeans as

> You are the promised kiss of springtime
> that makes the lonely winter seem long.
> You are the breathless hush of evening
> that trembles on the brink of a lovely song

In setting Jerome Kern's soaring melody for "All the Things You Are," Hammerstein exemplifies the traditional lyricist's art as defined by Anthony Burgess: his rhymes are simple, his imagery is unobtrusive, and his skillful manipulation of long vowels and verbal phrasing brings out the musical pattern and makes the lyric eminently "singable."

Cole Porter, however, could take the same "You are . . ." formula of romantic compliment and fashion a lyric so full of the "verbal shocks" of society verse, "so overtly clever and wittily brilliant" that, as Gerald Mast observes, "they overwhelm the music"[8]—a complete reversal of Burgess's formula for lyrical self-effacement:

> You're a rose,
> you're Inferno's Dante,
> you're the nose
> on the great Durante

In "You're the Top" the rhymes (*rose, nose,* Inferno's) are clever, the diction (right down to "you're" instead of Hammerstein's "You are . . .") casually colloquial, the sentiment flippantly antiromantic—all hallmarks of light verse. The images, moreover, are not only striking but demand the same concentration we give to poetry in order to see that what at first appears to be a clash of European classic and American vaudeville resolves into underlying harmony: superficially different, Dante and Durante both turn out to be Italian comedians, albeit the latter less divinely so.

Urbane yet casual, literate yet colloquial, sophisticated yet nonchalant, "You're the Top" (1934) epitomizes an era when song lyrics radiated the stylish verve of society verse. But the poets of Tin Pan Alley could also borrow features from the more avant-garde poetry emanating, a few blocks away, from Greenwich Village. The best Alley lyrics can be as ironically understated as a Millay sonnet or bristle with the linguistic by-play and foreplay of a cummings panegyric. Cleanth Brooks, one of the few literary critics to take even passing note of song lyrics, once observed that the same striking images that characterized modern poetry—Eliot's comparison of the evening "spread out against the sky" to "a patient etherised upon a table," for example—could also be found in such popular songs as "You're the Cream in My Coffee." Had Brooks looked further into that or numerous other lyrics of the day, he would have found other features he praised in modern poetry: wit, paradox, "ironical tenderness," and a "sense of novelty and freshness with old and familiar objects."[9]

In the wake of the famous Armory show of 1913, New York poets sought to adapt the techniques of such artistic movements as Cubism and Dadaism. Thus Marianne Moore could use the page as a canvas to arrange verbal fragments in rhyming shards:

>   ac-
>   cident—lack

In the same year, similarly, a lyricist like Lorenz Hart could use Richard Rodgers' music as a grid to break up words into equally clever rhyming fragments:

>   sweet pushcarts gently gli-
>   ding by

Just as e. e. cummings could construct a verbal collage where "Abraham Lincoln" was juxtaposed against an ad for "B. V. D" and "Lydia E. Pinkham" against "the girl with the Wrigley Eyes," Cole Porter's discordant list songs jarringly set "Botticelli" beside "Ovaltine," "Mahatma Gandhi" by "Napoleon Brandy."

When New York Dadaists began presenting ordinary "found" objects—a bicycle wheel, a snowshovel, a urinal—as "readymade" sculptures, poets like William Carlos Williams quickly followed suit, framing such prosaic objects as a red wheelbarrow or the "figure 5" in poetic lines. Poets also seized upon common verbal objects—ordinary catch-phrases like "so much depends upon" and "this is just to say"—and revealed their poetic qualities by fragmentation and juxtaposition.

Such "found" phrases were also the basis of many of the songs of the golden age, where the most banal colloquial idioms were lifted into the romantic space of a lyric. What could be less likely terms of endearment or heartache than these bits of vernacular junk: "I Guess I'll Have to Change My Plan," "I Can't Get Started," "What'll I Do?," "Sure Thing," "There'll Be Some Changes Made," "How About Me?," "How Long Has This Been Going On?," "It Never Entered My Mind," "It's All Right With Me," "Just One of Those Things," "You Took Advantage of Me," "I Didn't Know What Time It Was," "Say It With Music," "Say It Isn't So," "Don't Get Around Much Anymore," "I'm Beginning to See the Light," "You're Driving Me Crazy," "Ain't Misbehavin'," "I Should Care," "They Can't Take That Away From Me," "From This Moment On," "I Don't Stand a Ghost of a Chance," "Everything Happens to Me," "Day by Day," "Night and Day," "Day In—Day Out . . . ," a list that could go, as a more recent song has it, "On and On."

By the 1920s all of the arts emulated painting's emphasis upon its own medium, and the medium for Tin Pan Alley's lyricists, as it was for the poets of Greenwich Village, was what H. L. Mencken in 1919 pugnaciously dubbed *The American Language.* By adapting the techniques of modern poetry, as well as those of society verse, and wedding them to music, the lyricists of Tin Pan Alley took the American vernacular and made it sing.

Yet while much of their art resembles that of society verse and modern American poetry, much of it will be lost if we forget Ira Gershwin's warning. Since the lyricist's craft is one of fitting words "mosaically" to music, what we must notice is not only the witty image and cleverly fragmented rhyme but how well the lyricist works within musical constraints (or subverts them), how deftly he matches the composer's phrasing with verbal fits (or shrewd misfits). Seeing Ira Gershwin's lyric in print can reveal a clever pun like the one that closes "But Not for Me:"

> When ev'ry happy plot
> ends with the marriage knot—
> and there's no knot for me

But for other artful effects, we must read the words with the music in mind. Only then can we see why Ira loved the added restrictions of his brother's music, where abrupt, ragged phrases gave a lyricist little room to "turn around." In a song like "They All Laughed," for example, George Gershwin's melody starts out with a ten-note phrase, calling for a ten-syllable line,

> They all laughed at Christopher Columbus,

followed by a line of seven syllables:

> when he said the world was round.

Then the music seems to repeat the same pattern with another ten-syllable line,

> They all laughed when Edison recorded

But, instead of another seven-syllable line, it stops short on one note:

> sound.

Thus Ira fitted that tiny musical space with as much invention as the character he celebrated.

In *Lyrics on Several Occasions* Ira Gershwin traces the genealogy of this "mosaic" art back to the Elizabethan Age. However, he is careful to exclude the "great art-song writers," such as Dowland and Campion, since for them "the words always came first, even though many of these highly talented men were also fine composers and wrote their own lute accompaniments." Instead he looks to the "satirists and parodists" of the period "who, discarding the words of folk song and ballad, penned—quilled, if you like—new lyrics to the traditional tunes." This "practice of putting new words to pre-existent song" culminated in Gay's *The Beggar's Opera,* and continued a tradition that, for Gershwin, includes such unlikely precursors as Martin Luther, who put new, spiritual lyrics to old, worldly tunes. So severe is Gershwin in his insistence that the music must come first, he even excludes Gilbert and Sullivan operettas, since "practically all the lyrics were written first." Like a true purist, however, he includes Gilbert's earlier settings of music from Continental operas, cherishing their "tricky lyrics and recitatives loaded with puns."[10]

Gershwin's thumbnail survey stops just before the twentieth century, but it was then, with the emergence of Tin Pan Alley, that lyricists faced musical constraints more severe than ever before—or since. One of the earliest industries geared to standardization and mass marketing, Tin Pan Alley quickly evolved a rigid formula for popular songs. Almost every song from the golden age is built upon the same musical pattern of a thirty-two-bar chorus structured in four eight-bar units, usually in an AABA sequence. That is, the main part of the song consists of a melody, eight measures long, repeated three times with only another eight-bar phrase to vary it. The formula made songs easy to write: think of a pleasant eight-bar melody, repeat it, shift briefly to another melody for the eight-bar B section, or "release," then return to the main melody for the final A section. The formula was also a model of efficient repetition, playing a melody, repeating it, varying it briefly, then, knowing that all listeners, like Shakespeare's Duke Orsino, want "that strain again," returning to the original eight-bar phrase.

For the lyricist this tight musical pattern was extraordinarily restrictive, and many lyrics from the period simply follow the musical repetitions with banal, box-like phrases: "you were meant for me; I was meant for you." Such restrictions, however, could spark the inventiveness of a lyricist like Gershwin, who could cleverly make the syntax of his lyric spill over the musical boundaries. In "They Can't Take That Away From Me," for example, he concludes the musical release with a dangling preposition:

> We may never, never meet again
> on the bumpy road to love,
> still I'll always, always keep
> the mem'ry of—

Only in the final A section is the lyrical thought completed with "the way you hold your knife"—a wryly tender image that understates, even as it affirms, romantic feeling.

The musical pattern also put constraints upon the lyricist's subject matter. In nineteenth-century songs, as well as in most contemporary rock and country songs, the musical pattern is a sequence of verses, punctuated by simple eight- or sixteen-bar refrains, and the lyrics often recount stories about a wide range of subjects, from "The Convict and the Bird" to "The Little Lost Child." But the rigid thirty-two-bar AABA chorus of the golden age allows no such freedom. Such a "seamless web" made narrative, characterization, or social commentary practically out of the question. What it did allow, demand, in fact, was the expression of "one moment's feeling in a fluid statement" of between fifty and seventy-five words.[11] The feeling that lent itself best to such treatment was love—in, out of, or unrequited. Given such a musical formula, the problem for a lyricist was both simple and hard; as Doris Day puts it to Danny Thomas in the film biography of lyricist Gus Kahn, "Gus, ya gotta learn to say 'I love you' in thirty-two bars."

Little wonder, then, that between 1920 and 1940, 85 percent of popular songs were love songs—a substantially greater proportion than one finds in nineteenth-century songs or even in contemporary songs.[12] Prohibition, the Depression—these are barely mentioned in the popular lyrics of the golden age. Such a constraint of subject matter put Tin Pan Alley lyricists in the same straitjacket as their medieval ancestors, the troubadours of Provence, who, after all, invented this thing called romantic love. To the modern reader, who looks to poetry for original insight, sincerely expressed, popular song lyrics, like medieval *chansons,* "all sound the same, . . . sweet but bland repetitions of the few basic clichés of courtly love."[13] What such readers miss is the cleverness, the inventiveness and, in the best sense of the word, artifice, that displays itself by ringing endless changes upon what are indeed the tiredest clichés, the tirer the better for the skillful artificers of Provence. In the lyrics of Tin Pan Alley, similarly, we must listen, not for new ideas or deep emotion, but for the deftness with which the lyricist solves the problem posed by a song of the 1930s: "What Can You Say in a Love Song That Hasn't Been Said Before?"

The great lyricists solved this problem tirelessly. Lorenz Hart could describe falling in love with skeptical masochism:

> this can't be love because I feel so well—
> no sobs, no sorrows, no sighs

Ira Gershwin could have a newly smitten lover react to a first passionate kiss with an angry demand, "How Long Has This Been Going On?"—a catch-phrase we associate more with an irate spouse discovering a mate's infidelity. Cole Porter could terminate an affair with an urbane shrug—"it was just one of those things"—then add a metaphoric afterthought that contradicts the very ordinariness it asserts: "a trip to the moon on gossamer wings—just one of those things."

Hart, Gershwin, and Porter set an unsentimental, even anti-romantic standard for song lyrics, a standard that emerged practically overnight, when "Manhattan," a "sophisticated" song from a little revue called *Garrick's Gaieties* became an enormous popular hit in 1925. "Suddenly in the middle Twenties," librettist Sig Herzig recalled,

> I began to read about Larry [Hart] in the columns of the New York papers. . . . F. P. A., whose column, "The Conning Tower," was the arbiter of wit at the time, was mentioning Larry's lyrics, as were other columnists. It was a major breakthrough! Outside of a small coterie of Gilbert and Sullivan worshippers, a lyric was never noticed and the lyric writer remained anonymous. . . . Then other lyric writers started gaining attention and there was a gush of wonderful words and music[14]

No one was more amazed at the popular success of such a witty song than Rodgers and Hart themselves, who had struggled for years to establish themselves on Broadway. At the same time, Ira and George Gershwin began their successful collaboration with *Lady, Be Good!* (1924), and their songs, too, quickly detached themselves from shows to become independent hits on Tin Pan Alley's sheet music and record sales markets. By 1928, Cole Porter, who had been trying unsuccessfully to write popular songs for years, finally had both a hit show and a hit song, with the insouciant "Let's Do It" from *Paris.* Just as Porter's star began to rise, Oscar Hammerstein, who eschewed sophistication and maintained the old lyrical tradition of effacing words before music, saw his career go into decline.

For nearly two decades, Hart, Gershwin, and Porter's casually sophisticated standard was emulated by other lyricists. Theater lyricists could sidestep the elevated, melodramatic strains of operetta to nonchalantly lament,

> I guess I'll have to change my plan
> I should have realized there'd be another man!
> Why did I buy those blue pajamas
> before the big affair began?

Hollywood lyricists, too, could sometimes transcend the banal fare of most film songs:

> Thanks for the memory
> of rainy afternoons,
> swingy Harlem tunes,
> and motor trips
> and burning lips
> and burning toast and prunes.

Even lyricists who wrote for Tin Pan Alley's straight popular market could occasionally turn a sensuously witty compliment:

> You go to my head
> like a sip of sparkling Burgundy brew
> and I find the very mention of you
> like the kicker in a julep or two

Such lyrics are, in a word, urbane, and Charles Hamm is right to observe that in the golden age the style of American popular song became a "New York style":

> Even more than had been the case during the formative years of Tin Pan Alley, the field was dominated by composers and lyricists born and trained in New York, writing songs for publishers who not only had their offices in New York but were themselves products of the city. . . . There was little effective cultural input from the rest of America into New York in these days, and to the extent that Tin Pan Alley songs reflected American culture in a broader sense, they did so because the rest of the country was willing to accept a uniquely urban, New York product. . . . The songs of Kern, Gershwin, Porter and their contemporaries were urban, sophisticated, and stylish, and they were intended for people who could be described by one or more of these adjectives—or aspired to be.[15]

Thus listeners all over the country delighted in the urbane lyrics of Hart, Gershwin, and Porter with the same relish that "the little old lady in Dubuque" might await her copy of the *New Yorker* or a newspaper reader in Ohio might chuckle over the syndicated reports of Benchley and Parker witticisms at yesterday's luncheon at the Algonquin Round Table. What, after all, could be more deliciously ironic than the fact that Fred Astaire, the performer who most embodied that urbane style of casual elegance—and introduced many of the greatest songs of the golden age—was born in Omaha?

Such a homogeneous style, however, was bound to be short-lived. What Sig Herzig termed a "gush of wonderful words" crested in the late 1930s, then, with George Gershwin's death, Porter's crippling accident, and Hart's losing battle with alcoholism, it quickly began to subside. With World War II, sentimentality and nostalgia returned to displace nonchalant sophistication, and soon popular song began to reflect, as well as emanate from, other regions, races, and classes. Even in the heart of Manhattan the change was registered: when Richard Rodgers broke with Lorenz Hart, he turned to Oscar Hammerstein, whose lyrical style was the very antithesis of urbanity and sophistication. In 1943, when the curtain went up on *Oklahoma!*, an age that had begun with "Manhattan" was over.

Yet because the songs of the golden age set a new standard, not only in musical sophistication but in poetic artistry, they have continued to appeal to performers and audiences while songs from other eras have been relegated to period pieces. Many are as familiar today as when they were written, continually refreshed through interpretations by the most diverse performers in jazz, country, and rock idioms. Fittingly, we call such songs "standards"—though the poetic art that has helped them endure has, for too long, gone unsung.

After the Ball:
Early Alley

> *Look at newspapers for your story line; acquaint yourself*
> *with the style in vogue; know the copyright laws; avoid*
> *slang.*
>
> CHARLES K. HARRIS

In 1900 songwriter Monroe H. Rosenfeld was commissioned by
the *New York Herald* to do a story on the new sheet music publishing industry that had emerged at the end of the nineteenth
century. Rosenfeld went to the tiny stretch of West 28th Street
between Broadway and Sixth Avenue, where most of these publishers had their offices. There, out of the windows of the
closely packed buildings, came the din of dozens of upright
pianos—a din made even more tinny by weaving strips of newspaper among the piano strings to muffle the sound. The racket,
so the story goes, reminded Rosenfeld of rattling pans and inspired him to christen the street "Tin Pan Alley." Although the
location of Tin Pan Alley kept changing—it had started out in
Union Square, near Tony Pastor's vaudeville house, followed
the movement of theaters uptown at the turn of the century,
and followed them again in the 1920s to the stretch of Broadway between 42nd and 50th—whatever its address, it was always
located, as songwriter Irving Caesar once explained, "close to
the nearest buck."

It may seem surprising that the poetic lyrics of the 1920s and

19

'30s should emerge from such a crassly commercial industry where artistry, particularly in lyric writing, was never a concern. Yet it was precisely the Alley methods of "popularization" and "standardization" that created those constraints which, as we have seen, inspired the witty subversions of the great lyricists of the golden age.

The very notion of "popularizing" a song was practically invented by these new music publishers. Older, established firms dealt in classical piano pieces, choral church music, and music instruction booklets. If they published a "popular" song, it was usually after traveling minstrel companies or variety performers, like the "Singing Hutchinson Family," had created a demand for it. Even then, the demand was not heavy, since, as Warren Craig notes, "sheet music was found only in the nation's more affluent homes furnished with pianos."[1] Since the sale of 75,000 copies of a song such as Stephen Foster's "Massa's in de Cold, Cold Groun' " was considered "phenomenal" in 1852, it's not surprising that traditional publishers did not solicit or try to "popularize" songs:

> Composers, performers, even the public had to beat a path to their doors. To go out in search of song material, to manufacture songs for specific timely purposes or events, to find performers and even bribe them to introduce such songs, to devise ingenious strategy to get a public to buy the sheet music—all this was not in the philosophy of conducting a music-publishing venture.[2]

In the 1880s, however, mass production made pianos more affordable, and as these pianos graced more and more parlors, the demand for sheet music expanded enormously. To tap that market a new kind of music publisher began to emerge: for Maurice Shapiro and Louis Bernstein, Leo Feist, Edward Marks, Joseph Stern, and the Witmark brothers, songs were made, not born, "Made to Order," as one firm advertised its wares. Most of these new publishers were first- or second-generation Jewish immigrants. Many had started out as salesmen—Stern had sold neckties; Marks, notions and buttons; and Feist had been field manager of the R and G Corset Company, figuring, as Kenneth Kanter says, that "anyone who could sell corsets could also sell songs."[3]

It was in the marketing of their songs that these new publishers displayed their greatest creativity. What by the 1950s would be scandalously termed "Payola" began in the Alley as the perfectly respectable practice of bribing performers to sing your firm's songs. The bribe could range anywhere from buying dinner for a singer to paying her a royalty on its sheet music sales. Far more innovative than bribery, however, was the practice of "plugging" a song. In the publishing offices that lined 28th Street, piano "pluggers" relentlessly demonstrated their company's latest wares to the public as well as to professional entertainers in search of new material. (George Gershwin started out as a plugger for the Remick company—at age fourteen, the youngest piano player on the Alley). Pluggers also were sent out beyond the confines of Tin Pan Alley—hired to sing songs in restaurants, five-and-dime stores, street corners, beaches—anywhere a crowd was gathered, sometimes serenading them from a passing bicycle, sometimes hanging overhead from a balloon.

The best place to plug songs, however, was in the theater. By the late 1880s, the rise of vaudeville had displaced the minstrel show as the staple of the musical theater, and while songs in minstrel shows were performed with the entire company on stage, vaudeville numbers were done in cameo by individual performers, giving singers more control over what songs they sang. Getting a famous vaudeville star to make your song part of her act assured it of the widest possible dissemination, particularly by the early part of the century, when vaudeville was expanding across the country in a national "circuit." At its most inventive, plugging was done by a "singing stooge" (often a young boy with a good voice) planted in a theater audience. After a performer sang a song from the stage, the stooge would rise "spontaneously" and, as if carried away by the song's beauty, sing encore after encore until the audience joined in. (It was as just such a plugger, at Tony Pastor's vaudeville house, that Irving Berlin got his start in Tin Pan Alley.)

So effective were such methods that by 1900, as Charles Hamm notes, "control of the popular-song industry by these new publishers was virtually complete, and it was a rare song that achieved mass sales and nationwide popularity after being published elsewhere."[4] Reason enough to send out Monroe H. Rosenfeld to do a story on "Tin Pan Alley."

Mass marketing naturally called for mass production, which, in turn, meant standardization. Publishers devised musical and lyrical formulas for various kinds of songs so that any new song would sound instantly familiar to the public. Such formulas were so simple, moreover, that many of the most successful "composers," who could not read a note of music, simply whistled or hummed a formulaic melody to one of the publishing house staff arrangers, who then transcribed and harmonized it. Only afterward was a lyricist called in to put words to the melody, his title often dictated by the publisher, who kept his eye on the newspapers for topical subjects that could be tailored to the traditional formulas.

In these early days of Tin Pan Alley, however, songs were built on formulas less rigid than the " 'I love you' in thirty-two bars" model that constrained lyricists of the golden age. The formula for the sentimental ballad, the dominant kind of popular song in the 1890s, for example, differed radically from the "ballads" of the 1920s and '30s. For one thing, like most nineteenth-century songs, it truly was a *ballad,* a song that told a story, rather than the purely lyrical "ballads" of the golden age. The story, moreover, could be about virtually any subject, from "The Pardon Came Too Late" to "The Picture That Is Turned to the Wall." When it did deal with love, it was usually through a narrative as convoluted as a soap opera.

Charles K. Harris' "After the Ball" (1892) is typical of such "song-stories": a young girl at a ball asks her fiancé to fetch her a glass of water, but, when he returns to find her kissing another man, he drops the glass and storms off, only to learn, years later, that the man kissing her was her brother and that she died of a broken heart—all of this O. Henryish irony suitably framed as the fiancé, now a gray-haired bachelor "uncle," tells the story to his young niece! Such an absurd tale of misunderstanding was turned down by the first singer Harris urged to plug it; "If I sang a line like "Down fell the glass, pet, broken, that's all," she insisted, "the customers in my saloon would shatter their beer mugs in derision." Yet it was precisely this pathetic narrative that was marketed into the first big hit from Tin Pan Alley, selling over five million copies of sheet music and galvanizing the new publishing industry by revealing how much money could actually be made from a popular song.

Critics who complain that the lyric is merely a "thoroughly artificial piece of doggerel"[5] forget that in the sentimental ballad it was the music and the story—rather than the poetic texture of the lyric—that counted. In an extensive analysis of those elements, Mark Booth has shown why "After the Ball" became so successful; the music, he argues, has "considerable merit" and the story is "easily recognizable as the property of the English and American Victorian culture from which it comes. It is the substance of innumerable melodramas and sentimental novels . . . condensed and poised in a few bare lines."[6]

Telling such a story and fitting it to music frequently forced a lyricist like Harris into padding ("List to the story—I'll tell it all"), awkward inversions ("Where she is now pet, you will soon know"), and mismatches of musical and verbal accent (so that "*ball*room" must be pronounced "ball*room*"). But if the narrative line worked, the verbal manner didn't matter. Harris waved away all such infelicitous lyrics as the result of "certain allowances" that had to be made to fit words to music. "When the song is rendered," he insisted, the "defects are not so apparent."[7]

Since he frequently had to make such stylistic contortions to set his story to music, the lyricist usually adopted an elevated "poetic" diction that allowed for inversions and elisions. Not only did such diction make storytelling easier, it heightened the melodrama of "song-stories" where chaste maidens face their tormentors "with cheeks now burning red." The last quality one expects to find in a sentimental ballad is vernacular ease, and Harris himself was a staunch advocate of the elevated style, advising would-be songwriters to "avoid slang."

Storytelling was tied to a musical formula, derived from nineteenth-century songs, that consisted of a series of regular "strophes" or verses, each verse followed by the same chorus (or, as it was called then, the refrain). It was in the verses that the story unfolded, and the chorus merely echoed or commented upon it. As Charles Hamm has noted, however, the relationship between verse and chorus gradually shifted so that, by the end of the nineteenth century, songs had fewer and fewer verses and the "chief melodic material" had shifted to the chorus.[8] "After the Ball," in fact, strikes an even balance between three verses and three refrains. One measure of how that shift continued throughout the history of Tin Pan Alley is that anyone who knows "After the Ball" today probably knows only its refrain.

It is in that refrain that we can begin to see the lyrical lineaments of the standard chorus of the golden age. Unlike the eight- or sixteen-bar refrains of most nineteenth-century songs, Harris' is thirty-two bars, but rather than dividing those thirty-two bars into separate segments, as later lyricists would do, Harris' lyric is all of a piece—one extended sentence that begins with four parallel phrases:

> After the ball is over,
> after the break of morn—
> after the dancers' leaving;
> after the stars are gone;

The parallelism gives formal weight to the sentence and, as the music rises to a climax, Harris reaches his main clause:

> many a heart is aching,

But then the melody climbs higher, traversing an entire octave and even an interval beyond, and the lyric compounds the sadness—not only do they ache but they ache, alas, beyond our ken!—in a soaring subordinate clause,

> if you could read them all;

As the music descends to its close, Harris shifts from clause to phrase, concluding with what became a staple Alley device for plugging a song—repeating its title at the end of the chorus:

> many the hopes that have vanished
> after the ball

Strained though it seems, the refrain offered Harris more lyrical freedom than his verses. Since he did not have to tell a story, he could use the thirty-two bars to develop a simple emotional lament with considerably less melodrama and "poetic" diction than he used in the verse. Nevertheless, Harris, like other early songwriters, regarded the lyrical formula for the sentimental ballad as one of strophic storytelling, with the refrain still serving, as in nineteenth-century song, as commentary and echo on the verses. He persisted in grinding out lachrymose "song-stories" for years: "Just Behind the Times" (1896) bewailed the forced retirement of an old-fashioned minister; "Break the News to Mother" (1897) were the dying words of a Civil War soldier given new life (the song, not the soldier) by the Spanish-American War; "Hello, Central, Give Me Heaven" (1901) outdid all other "telephone songs" with its agonizing portrayal of a little girl who vainly tries—and *keeps* trying—to call her dead mother on the newfangled instrument. Mired in the "song-story" formula, the sentimental ballad was a far cry from the urbane and vernacular love songs of the golden age. Before it could be reborn, the sentimental ballad, alas, would have to die.

What killed it was ragtime. Ever alert to shifts in musical taste, Tin Pan Alley publishers noted the increasing popularity of syncopated music and quickly began adapting the flavor of ragtime into a new song formula that temporarily displaced the sentimental ballad as the industry's major product. Ragtime entered Tin

Pan Alley in the late 1890s and merged with "coon" songs—
syncopated (though sometimes only barely so) comedy songs derived from minstrel shows. As in the minstrel show, they were
performed in vaudeville by whites in blackface. Some of the ragtime "coon" song composers were black, such as Ernest Hogan,
who lived to regret having written such numbers as "All Coons
Look Alike to Me" (1896), but most were whites who turned out
racist caricatures in confected black slang.

Trying to write in slang—and fit words to even mildly syncopated music—called for a different lyrical formula. Few ragtime
"coon" songs, for example, tell a story; most sketch a situation,
usually in one or two brief verses, then concentrate on the chorus, where a caricatured black might plead "I Want Yer, Ma
Honey!" (1895), celebrate "My Black Baby Mine" (1896), or lament "You've Been a Good Old Wagon But You Done Broke
Down" (1896). Not only did such songs create a shift from narrative to more "lyrical" lyrics, the use of dialect made the words
themselves more playfully prominent. In Harney's "Mister Johnson, Turn Me Loose" (1896), the first ragtime song to become a
major hit, for example, listeners were delighted not only by the
syncopated rhythm but by the idiomatic plea of the caricatured
black to "Mister Johnson" (slang for the police):

> Oh, Mister Johnson, turn me loose!
> Don't take me to the calaboose!

At first, the lyrics of the "coon" song "were as different from the
words of the waltz-tragedies as was the music of those waltzes
from the jagged melodies of the raging 'rags.' " When those lyrics turned to love, as they increasingly did, they created what
Isaac Goldberg terms "a vocabulary of unadorned passion—a
crude *ars amandi.*"[9] Thus with a forthrightness unimaginable in a
Harris ballad, a coon-shouter could plead, "All I want is lovin'—I
don't want your money."

No sooner had the ragtime "coon" song established itself on
the Alley, however, than lyricists began mingling its vernacular
idioms with the elevated diction of the sentimental ballad. Such

stylistic clashes as "yon nig" color Hogan's lyrics, and in Barney Fagan's "My Gal Is a High Born Lady" (1896), a black bridegroom looks forward to his wedding day thus:

> Sunny Africa's Four Hundred's gwine to be thar,
> to do honor to my lovely fiancee;
> thar' will be a grand ovation
> of especial ostentation

Kerry Mills' "At a Georgia Camp Meeting" (1897) shows the same stylistic schizophrenia, suddenly switching from "how the 'Sisters' did shout" to " 'Twas so entrancing." Indeed, by the turn of the century, one might say of the "coon" song the same thing one of its caricatured lovers says of his "baby": "She's Getting Mo' Like the White Folks Every Day" (1901):

> Now she can sing "The Swanee River"
> like it was never sung before,
> but since she's worked in that hotel
> she warbles "Il Trovatore"

It was, in fact, a team of black songwriters who nurtured the hybrid of coon song and sentimental ballad. By making its lyrics "noticeably more genteel," J. Rosamond Johnson, his brother James Weldon Johnson, and Bob Cole were hailed as the collective "Moses" who led "the coon song into the promised land."[10] In songs like "Under the Bamboo Tree," the writers, according to Rosamond Johnson himself, tried to "clean up the caricature," using only "mild dialect" to express love "in phrases universal enough" to meet the "genteel demands of middle-class America."

It was lyricist Bob Cole who suggested to Johnson that "Nobody Knows the Trouble I've Seen" could be turned into a ragtime song. Johnson at first thought the suggestion sacrilegious, but at Cole's insistence, he syncopated the spiritual. In the lyric, Cole placed his lovers in Africa, where he could sidestep the coon caricature by having a dignified "Zulu from Matabooloo" propose marriage to a demure "maid of royal blood though dusky shade." Cole came up with what Sigmund Spaeth calls "a brand

new synthetic dialect"[11] and used what would become Tin Pan
Alley's favorite grammatical error—using *like* as a conjunction:

> If you lak-a me,
> lak I lak-a you,
> and we lak-a both the same,
> I lak-a say
> this very day
> I lak-a change your name

Here, too, one can see another feature of ragtime songs unheard
of in the sentimental ballad—deliberately using the music to dis-
tort or "rag" the lyric. The "lak-a" gave Cole a verbal equivalent
for the syncopated eighth-note/sixteenth-note pattern and inspired him to create something no "song-story" lyricist would
ever bother with—a pun, and a triple pun at that, on "like" as
wish, as *love,* and as *as.* Such "ragging" of words by music—
reversing verbal accents, breaking up phrases, splitting words by
musical pauses and rhythmic shifts—as well as the verbal play it
inspired, would become prominent features in the lyrics of
Lorenz Hart, Ira Gershwin, and other songwriters of the golden
age.

Whether or not "Under the Bamboo Tree" cleaned up—or
merely transplanted—the racist caricature, it certainly made for
a more artful lyric, one that achieved its effects not by storytelling but by a witty fit (and sometimes ragged misfit) between
verbal and musical phrasing. Its colloquial simplicity, moreover,
offered a new idiom for romantic songs—a cut above the dialect
of the "coon" song yet still well below the highflown style of such
ponderous narratives as "A Bird in a Gilded Cage." The ragtime
"coon" song, as Max Morath argues, had begun "transcending its
racial slur and dialect" and "licensing the use of slang and collo-
quialism, even bad grammar"[12] in all popular songs.

Some of those songs barely sound like "coon" songs at all;
only if one listens carefully to the words of the verse (or sees the
original sheet music, adorned with grotesque racist caricatures),
does their heritage emerge. Yet their vernacular idiom, their

comic touches, their passionate flair breathed fresh air into Tin Pan Alley. "Hello, Ma Baby" (1899), for example, refreshes the standard telephone greeting by "ragging"—reversing—the verbal accent against the musical beat: not the normal "Hel-*lo*" but "*Hel*-lo, my baby." Equally refreshed are the clichés of romance when they are couched in slang that permits more exuberant expressions of love than ordinarily permitted to demure white lovers:

> Send me a kiss by wire,
> Baby my heart's on fire!

Comedy and slang also permitted a wry "coon" lament in "Bill Bailey, Won't You Please Come Home?" (1902), where the musical accents disrupt the verbal syntax and leave it momentarily dangling.

> 'member dat
> rainy eve dat
> I drove you out,
> wid nothing but a fine tooth comb?

While the image betrays the lyric as a "coon" song (what could be more useless than a fine tooth comb for the wooly-haired caricatures depicted on the sheet music?), it provides a flash of verbal wit as *comb* brings out the clever ear and eye rhymes in the title phrase: *come* and *home*. Such an anguished plea in wry slang would naturally be unthinkable in a sentimental ballad, but it would not be long before such passion—and humor—worked their way into mainstream lyrics, along with the brassy colloquial style that made such feelings articulate.

The key figure in the outward spread of ragtime syncopation and vernacular lyrics to Broadway was George M. Cohan. Although his earliest songs, such as "When the Girl You Love Is Many Miles Away" (1893), were in the elevated, sentimental strain, he quickly turned to "coon" songs. Learning that the "coon" song code word for sex was "heat"—"Dar's No Coon Warm Enough For Me," "A Hot Coon From Memphis," "She's My Warm Baby," "A Red Hot Coon," "The Warmest Colored Gal in Town"—the teenaged Cohan quickly added his own "The Warmest Baby in the Bunch" (1896). He then shrewdly reversed the formula with "You're Growing Cold, Cold, Cold," advertised as "the story of a coon with an iceberg heart." Contributing to the glut of "coon" technology songs with "I Guess I'll Have to Telegraph My Baby" (1898), Cohan used slang even more coyly:

> For Lucy is a very gen'rous lady,
> I can always touch her for a few.

Already here he displayed his knack for subtly simple rhymes— *have*/telegra*ph*, *Lucy*/*few*.

By the beginning of the twentieth century Cohan had established himself on Broadway and begun his war to rid the American musical stage of its dependence upon European models— Viennese operetta, French *opéra bouffe*, and the British comic opera of Gilbert and Sullivan. Having absorbed the syncopated music and vernacular idiom of ragtime, he used them to give Broadway a homegrown alternative to its usual diet of Viennese schmaltz and English patter. While *Little Johnny Jones* (1904) is now regarded as the first genuine American musical, it fared poorly on the stage. Undaunted, Cohan peddled its songs directly on Tin Pan Alley and turned them into independent hits:

> Give my regards to Broadway,
> remember me to Herald Square,
> tell all the gang at
> Forty-Second Street, that
> I will soon be there

Here, with the ragtime device of matching short and long vowels to the eighth-note quarter-note pattern, Cohan's handling of colloquial formulas and place names seems effortless. He also uses musical phrasing to truncate his lyric and create a mosaic of such "ragged" fragments as "Forty-second Street that" that create up-beat rhymes on "*that*" and "*at.*"

To give voice to such lyrical pyrotechnics, Cohan created a character—an insouciant, nonchalantly sophisticated New Yorker, as urbane as the city he celebrated. In Cohan's hands the energetic dialect of the caricatured "coon" metamorphosed into the vernacular ease of a jaunty cosmopolite who could even salute the flag in slang, though complaints from several patriotic societies forced him to revise his original line, "You're a grand old rag" to

> You're a grand old flag,
> you're a high-flyin' flag

Fragmenting the tiredest clichés into fresh, hard-edged units, such a Yankee sidestepped sentimental patriotism through brusque understatement:

> You're the emblem of
> the land I love,
> the home of
> the free and
> the brave

Not only do these abrupt rhythmic accents fragment the verbal line, they disclose internal rhymes on "*land*" and "*and*," "*em*" and "*blem*." Cohan's flippant "of"/"love" rhyme (a desperate maneuver in a language that gives a lyricist fewer rhymes for "love" than for almost any other four-letter word) turns up again in the slangily antiromantic lovers of the golden age, who lament "I can't give you anything but love, baby—that's the only thing I've plenty of, baby."

For years, however, Cohan was a lone vernacular voice on Broadway. At the beginning of the twentieth century, Franz Lehár, Rudolf Friml, and Victor Herbert "led the American public," as Leonard Bernstein puts it, "straight into the arms of operetta" with its exotic settings, melodramatic plots, and lush music. The lyrics for such operettas introduced the principle of "integration" into the musical theater—tailoring a lyric to character and story. Nevertheless, some of the songs, such as Herbert's "Ah! Sweet Mystery of Life" (1910), detached themselves from their theatrical context to become independently popular. What Bernstein describes as their "stilted and overelegant" lyrics

found a narrow niche in Tin Pan Alley, one that had been carved
out as early as 1889 with "Oh Promise Me" and that was filled
with such high-pitched sonorities as "I Love You Truly" (1901)
and the even more soaring platitudes of "Because (you come to
me with naught save love)" (1902).

"Meantime, just across the street," as Bernstein puts it, "the
vernacular was booming away on its own in the plotless musical
theater"—vaudeville, burlesque, and the revue.[13] Here the whit-
ening of the "coon" song continued, as "coon-shouters" metamorphosed into brassy singers like Eva Tanguay, who became known
as the " 'I Don't Care' Girl" for her rendition of Jean Lenox's lyric
to a syncopated song of 1905 by Harry Sutton:

> They say I'm crazy,
> got no sense,————————
> but I don't care!
> They may or may
> not mean offense,
> but I don't care!

As such singers began touring the country through a spreading
network of theaters, vaudeville quickly emerged as Tin Pan Alley's most lucrative avenue for plugging a song.

Vaudeville spawned numerous offspring, such as the ex-
travaganza and the "follies," the most prominent of which,
Ziegfeld's, began its annual series of lavish productions in 1907.
Since songs for such productions did not have to be integrated
into character and story, a singer like Nora Bayes could make her
sensational debut with a song redolent of slangy "coon" lyrics:

> I ain't had no lovin' since
> April, January, June or July

Vaudeville and its offshoots remained the major home of what
Bernstein calls the "musical vernacular," as well as of ragtime
songs and colloquial lyrics, until the 1920s.

If one single song from vaudeville may be said to mark the
complete integration of the "coon" song and the romantic ballad,

it would be "Some of These Days" (1910), a song which Alec Wilder finds, musically, to be a "landmark in popular music, perhaps *the* landmark song" of the early Alley."[14] It was written, in fact, by a black songwriter, Shelton Brooks, who managed to place it with Sophie Tucker, with whom it became indelibly associated. Tucker, in turn, had been a blackface "coon shouter" in vaudeville, but one night, so the story goes, she lost her makeup and costumes and had to perform as a white. Finding she could still hold an audience, she declared "I'm through with blackface."

The lyric of "Some of These Days" seems to be the "coon" song's way of saying it, too, was through with blackface. It inverts the comically pathetic lament of "Bill Bailey, Won't You Please Come Home?" and gives it a sophisticated pugnacity:

> Some of these days
> you'll miss me, honey,
> some of these days
> you'll feel so lonely

Here, with its easy contractions, its off-rhymes on "honey" and "lonely," was a thoroughly colloquial lament, right down to the oddly idiomatic catch-phrase title—not "one of these days," but the slangily skewed "some of these days."

Soon the hybrids of the "coon" song and the sentimental ballad became songs with the timeless character of a standard—songs that sound as if they could have been written anytime between 1910 and 1950. Many people are surprised to learn, for example, that "You Made Me Love You" was written as long ago as 1913, and are more surprised still to learn that Joseph McCarthy and Jimmy Monaco wrote it for Al Jolson's blackface routine at the National Winter Garden. Jolson had made his Broadway debut in 1910 in blackface, performing a "coon" song, "Paris Is a Paradise for Coons" (by none other than Jerome Kern), but it was "You Made Me Love You" that inspired his patented knee-drop on "gimmee, gimmee what I cry for" (though, legend goes, it was an ingrown toenail that downed him).

The lyric is an artful blend of the "coon" song and the sentimental ballad. McCarthy took the street taunt "You made me . . ." to enliven the clichéd lament of a lover trapped by fate, then suddenly dropped that bitter accusation to offer an equally vernacular, but limp, apology—"I didn't want to do it"—a juxtaposition of idioms that portrays the singer as at once helpless and aggressive. Picking up on the other meaning of "want" (as desire), McCarthy intensifies the paradox of passionate impotence with comically angry accusations:

> You made me want you,
> and all the time you knew it
> I guess you always knew it

Phrases oscillate between the helpless "cry for" and the street bully's "Give me" (which Jolson soon began delivering as "Gimmee, gimmee"). Similarly, in

> You know you got
> the brand of kisses
> that I'd die for

McCarthy juxtaposes the advertising jargon of "brand of kisses" against the romantic formula "that I'd die for." Such jagged fragments make the lyric a slang collage, framed by the intensified street taunt "*You know* you made me love you!"

Although initially done as a "coon" song, "You Made me Love You" became a staple in the repertoire of singers such as Fanny Brice and Ruth Etting, as well as the big bands of the 1930s. Even as late as 1946, a recording by Harry James could sell a million copies—one sure hallmark of a "standard." The song was also interpolated into numerous films and even launched Judy Garland's career when she sang it to a photograph of Clark Gable in *The Broadway Melody of 1938*. The Hollywood studios had to supply a new verse for the little girl, beginning "Dear Mr. Gable . . . ," and had to bowdlerize such lines as "Give me, give me what I cry for" with such pale substitutes as "The very mention of your name sends my heart reeling." Even with Garland's cooing, however, McCarthy's lyric still breathed some of its erotic anger.

Just as the ragtime "coon" song was being fully integrated into romantic ballads, a more potent strain of black music, the blues, began to make its foray into Tin Pan Alley. While W. C. Handy persisted throughout his life in the absurd claim that he "invented" the blues, he certainly can at least be credited with transforming blues into popular songs. Handy had started out trying to write straight Tin Pan Alley songs, but having absorbed blues from country bands in the South, he found that "folks would pay money for it." His first successful blues was actually a political campaign song for Mayor Crump of Memphis, designed to appeal to the black vote—which it did—in 1909. Four years later Handy sold the song to an Alley publisher for fifty dollars, a lyric was added, and, as "The Memphis Blues," it became the first commercial blues song to become popular, though Handy received none of the profits.

Despite the popularity of "Memphis Blues," Handy had great difficulty selling his next song, "St. Louis Blues" (1914). Turned down by every Alley publisher, he finally set up his own publishing company to issue the song. Still it did not become popular for years, until vaudeville and revue singers began using it in their acts. One impediment to its popularity was the fact that, just as the thirty-two-bar AABA chorus was becoming the standard formula for popular song, "St. Louis Blues" is primarily based on the twelve-bar AAB pattern of authentic blues:

> I hate to see de ev'-nin' sun go down, (A)
> hate to see de ev'-nin' sun go down. (A)
> Cause ma baby, he done lef' dis town (B)

That musical structure, like a poetic couplet, allows for witty contrasts, turns, and juxtapositions between the repeated A line and the final B line of each verse.

The middle of "St. Louis Blues," however, is, according to Alec Wilder, "unlike anything in any blues, work song, country song,"[15] and lyrically it is more like a popular song refrain, albeit one laced with more truncated slang, telegraphic syntax, and

sharp imagery than would appear in popular lyrics until the
1920s:

> St. Louis woman
> wid her diamon' rings
> pulls dat man roun'
> by her apron strings.
> 'Twan't for powder
> an' for store bought hair
> de man I love
> would not gone nowhere

While it is clearly a song about blacks—the St. Louis woman is a
black woman who wears a wig and powder to "whiten" herself—
there is none of the comic distance of the "coon" song. The
lament is alleviated only by the singer's own wry metaphors:

> I loves dat man lak a schoolboy loves his pie,
> lak a Kentucky Col'nel loves his mint an' rye

Such metaphors, moreover, testify to her sensuous delight in the
very blackness that the "St. Louis woman" covers up:

> Blackest man in de whole St. Louis.
> Blacker de berry, sweeter is the juice.

Ironically, while she delights in her lover's blackness, he desires
the white wig and powder of her rival.

By the early 1920s, when "St. Louis Blues" finally became
popular, features of the blues began appearing in more and
more commercial popular songs. "Classic" blues recordings by
Ma Rainey and Bessie Smith were sold as "race records" marketed directly for a black audience, but Tin Pan Alley publishers
soon began calling for songs with the flavor—sometimes only a
dash—of the blues. Most popular songs that call themselves
blues, such as "Am I Blue" or "Birth of the Blues," are really built
on the thirty-two-bar Alley standard formula—like that St. Louis
woman herself, whitened versions of black blues.

Still, the blues exerted a powerful influence on all song lyrics during the golden age, anchoring them in a sensuously vernacular idiom. The blues also, as Arnold Shaw has pointed out, increased the emphasis upon lyrics "exploring many sides of the loss of and longing for love."[16] The "torch song" of unrequited love, he notes, is "sometimes described as a white offshoot of the blues," emerging clearly in such hits as "I Ain't Got Nobody" (1915) and "After You've Gone" (1917), coloring the great "ballads" of the 1920s, from Berlin's "All by Myself" to Richard Whiting's "(I Got a Woman Crazy for Me) She's Funny That Way" (1928), and flowering in the earthy lyrics Johnny Mercer set to Harold Arlen's bluesy melodies in the twilight of the golden age.

Not only was 1914 the year the blues moved up to Tin Pan Alley, it was also the year that Broadway, the last bastion of the elevated style, finally began succumbing to the force of the vernacular. "American taste in theater music," Gerald Mast points out, finally began to desert operetta "for a more informal, colloquial American sound."[17] While the operettas of Sigmund Romberg, Rudolf Friml, and Victor Herbert would continue to spout Viennese schmaltz well into the 1920s, in 1914 George M. Cohan's long-standing campaign to Americanize the American musical found a major ally in Jerome Kern. Kern's music was rooted in the European tradition, but, while it often pulled a lyric toward sonorous heights, it could also sparkle with a simple, yet sophisticated, grace.

Although Kern had been writing music since 1902, he had his first major hit with an interpolated song for *The Girl From Utah*, "They Didn't Believe Me" (1914), a song that nearly equaled the international popularity of "Alexander's Ragtime Band." The success of "They Didn't Believe Me" confirmed the wisdom of Kern's hesitant shift from nineteenth-century musical formulas to the Alley's thirty-two-bar standard. It also, according to Martin Gottfried, "invented stage ballads"—theater songs modeled on the Alley formula so that they could detach themselves from their theatrical context and become independently popular songs—albeit "pop music of a decidedly higher class."[18]

In "They Didn't Believe Me," Kern's casually sophisticated melody was set to equally urbane words by Michael E. Rourke (using the pen name Herbert Reynolds). Even the verse sounds like spontaneous conversation:

> Don't know how it happen'd quite,
> may have been the summer night.
> May have been, well, who can say!
> Things just happen anyway.

The chorus opens, even more off-handedly, with "And," and in the initial section Rourke strings together a sequence of phrases that, even repeated, sound perfectly natural:

> And when I told them
> how beautiful you are
> they didn't believe me!
> They didn't believe me!

Only in the middle section of this ABA song does Rourke strain as he tries to follow Kern's shifts of key and rhythm with a quick inventory (your lips, your eyes, your cheeks, your hair). Although he fights to stay colloquial with "are in a class beyond compare," he slides into the elevated "You're the loveliest girl that one could see."

The problems widen in the second half of the song, musically an extended reworking of the first A section, with its notorious triplet. It was supposedly Kern himself who matched the triplet with the clumsy phrase,

> and I *cert-n'ly am* goin' to tell them

requiring a singer to sing "cert-n'ly" as two syllables (most singers have since solved the problem for themselves with a simple contraction: "And I'm *cer-tain-ly* going to tell them"). While Kern may have been the triplet culprit, Rourke fumbled along with such awkward inversions as "That I'm the man whose wife one

day you'll be." He did, however, manage a graceful modulation
of the title phrase at the conclusion:

> They'll never believe me,
> they'll never believe me

If anything, this catch-phrase is even more conversational than
the title, and the contraction "they'll" rhymes subtly with "And
when I *tell* them."

Kern moved Broadway still further away from European
opulence toward streamlined, American nonchalance with a series of musicals at the Princess Theatre between 1915 and 1918.
Since the Princess Theatre had only 299 seats, musical productions had to be scaled down radically from the florid extravaganzas exhibited by Ziegfeld and the Shuberts. "These shows," Da
vid Ewen explains, "helped to introduce a new kind of musical
theater into Broadway, consisting of intimate entertainment,
small casts and orchestra, economical scenery and costuming, an
intimate tone, an informal manner, and a bright, sophisticated
air."[19]

In the Princess shows, Kern's casually elegant music found
its perfect complement in the lyrics of P. G. Wodehouse. While
Wodehouse had his roots in the witty patter of Gilbert and Sullivan, he adopted what for him was an unorthodox practice in
setting Kern's melodies.

> Jerry generally does the melody first and I put words to it. W. S.
> ("Savoy Operas") Gilbert always said that a lyrist can't do de-
> cent stuff that way, but I don't agree with him, not as far as I'm
> concerned, anyway. If I write a lyric without having to fit it to a
> tune, I always make it too much like a set of light verse, much
> too regular in meter. I think you get the best results by giving
> the composer his head and having the lyrist follow him. For
> instance, the refrain of one of the songs in *Oh Boy* began, "If
> everyday you bring her diamonds and pearls on a string." I
> couldn't have thought of that, if I had done the lyric first, in a
> million years. Why, dash it, it doesn't *scan*. But Jerry's melody
> started off with a lot of twiddly little notes, the first thing em-

phasized being the "di" of "diamonds," and I just tagged along after him. . . . Anyway, that's how I like working, and to hell with anyone who says I oughtn't to.[20]

Although Gilbert may have set a model for witty elegance, the fact that his lyrics preceded Sullivan's music makes them scan like society verse:

> When I merely from him parted,
>   We were nearly broken-hearted;
> When in sequel reunited,
>   We were equally delighted.

What Wodehouse had found—and revealed to young lyricists (and would-be writers of *vers de société*)—is that one could turn the liabilities of popular song writing into an asset—by letting the music come first one could rhyme in equally clever ways without departing from colloquial diction and conversational phrasing:

> What bad luck! It's
> coming down in buckets

Here the rhyme emerges by fragmenting the syntactic phrase against the musical one—an elegant form of "ragging." What's more, Wodehouse's lyric sounds thoroughly colloquial and conversational, where Gilbert's sounds "poetic."

Thus it was Wodehouse, rather than Gilbert (whom Ira Gershwin, we recall, had excluded from his genealogy of lyricists), who showed how intricately words could be fitted "mosaically" to music "already composed." It was the very constraints imposed by such music, in fact, that sparked his inventiveness—using an unusual musical accent to highlight a rhyme between "that you" and "statue," grasping Kern's tricky insertion of three extra notes as a chance to turn a conversational phrase that debunks sentimentality: "I love him because he's—*I-don't-know*—because he's just my Bill."

Although the Princess shows ended in 1918, Wodehouse's lyrics inspired young lyricists in the same way that Kern's music revealed to young George Gershwin that popular songs could transcend the usual ricky-tick fare of Tin Pan Alley's tunesmiths. The key to such transcendence, as Gershwin himself realized, was to have his popular songs emanate not from the windows of Tin Pan Alley's publishing houses, where he had begun working as a piano plugger, but, like those of Kern and Wodehouse, from the more sophisticated reaches of the Broadway theater.

Tin Pan Alley had known from the beginning that the American musical theater was the best place to huck its wares. "After the Ball" for example, was simply interpolated into the show *A Trip to Chinatown*, with no regard whatsoever to its relation to characters, story, or setting. But with the invasion of European operetta in the early 1900s, as we have seen, came the integrated song, and while occasionally one of these became independently popular, Tin Pan Alley publishers contented themselves with interpolating their songs into vaudeville, burlesque, extravaganzas, "follies," and other loose variety shows. By 1920, however, both vaudeville and operetta began to wane, and the growing popularity of more thoroughly American musical comedies made Broadway "a glittering song supermarket, an easy way for Tin Pan Alley to plug its material."[21]

To be sure, Tin Pan Alley's influence on Broadway impeded the growth of the "integrated" musical, and historians of the musical theater understandably decry that influence. "The whole growth of our musical comedy," argues Leonard Bernstein, "can be seen through the growth of integration," which "demands that a song come out of the situation in the story and make sense with the given characters."[22] With a few exceptions such as *Show Boat* and *Of Thee I Sing*, "integration" did not become a watchword on Broadway until Rodgers and Hammerstein's *Oklahoma!* in 1943. Until then, the "book" of a musical was inconsequential fluff; a "good" musical was simply one with a lot of hit songs. Those songs were not only written with little regard for the story, they were frequently shunted in one show and out of the other. ("The Man I Love," for example, was yanked in and out of three different musicals.) After *Oklahoma!*, however, more and more musicals strove for integrated songs, and with the growth of integration fewer and fewer songs from musicals have become popular. Today, for example, it is a rarity for a Broadway show, even by such a successful composer and lyricist as Stephen Sondheim, to generate a hit.

What was bad for the growth of the American musical, however, was wonderful for the development of the popular song, for it meant that composers and lyricists writing for the theater in the 1920s and '30s also aimed for Tin Pan Alley's popular market. On the one hand, such songs had to have a measure of wit and sophistication, yet they also had to have some of the sensuous, vernacular ease that had evolved from the Alley hybrid of sentimental ballad and ragtime "coon" song, an ease further called for by the growing popularity of the blues. Songs most likely to succeed were published, recorded, and plugged over radio—a vehicle for "popularizing" songs that quickly displaced older Alley methods. Thus the great "unintegrated" lyrics of Lorenz Hart, Ira Gershwin, and Cole Porter were heard and bought by people who not only had not seen the musicals they came from but probably had never even heard of the shows.

Virtually all of those songs subscribed to the Alley formula of saying "I love you" in thirty-two bars, and thus for more than twenty years, American popular songs, whether they emerged from Broadway, Hollywood, or an Alley publishing house, whether popularized over the radio, by a dance band, or a singer's recording, all sound fundamentally alike, lyrically as well as musically. Charles Hamm attributes such homogeneity, during a period that saw "the most astounding technological changes yet experienced since popular song became part of American life," to two factors. The first was the founding, in the crucial year of 1914, of ASCAP (the American Society of Composers, Authors and Publishers). ASCAP was originally organized to insure that, before songs written by members of the organization could be performed publicly (or, later, played over the radio), fees were paid that ASCAP then distributed to copyright holders. The organization quickly grew to include almost all composers and lyricists and thus became a monopoly "whose chief concern, quite naturally, was with the type of music already being produced, rather than with new types of music perhaps more appropriate" to such new media as the phonograph record, radio, the sound movie. Such "standardization" meant that "the songs performed on radio and in the movies were written in a style born in vaudeville and other forms of musical theater in the late nineteenth and early twentieth centuries." Thus "there is no way to tell, from listening to a song by Irving Berlin or any of his contemporaries, whether it was written for vaudeville, musical comedy, the movies, or simply composed for radio play and possibly recording."[23]

Yet Hamm goes on to point out that an equally important reason for the "persistence of a single musical style in the popular songs of the Tin Pan Alley era is simply that this style was a vital, viable, successful, somewhat flexible, and relatively new one." Saying "I love you" in thirty-two bars remained a constraining but inspiring formula for lyricists for nearly three decades, its pattern of musical repetition and variation inviting deft turns of emotion, from straightforward sentiment to flippant irony, from grandiosity to self-deprecation, from lamentation to insouciance.

Joseph McCarthy, to take just one example, was clearly thinking in terms of this standard structure by 1918, when he wrote "I'm Always Chasing Rainbows" and interpolated it into the Broadway show *Oh, Look!*. Since composer Harry Carroll took his melody from Chopin's *Fantaisie Impromptu,* one might think McCarthy would have chosen a commensurately elevated style; instead he took a catch-phrase title, rich in long vowel sounds, and followed it with a perfectly conversational lament that unfolds across the four eight-bar musical sections, marking his transitions from section to section with shifts of rhyme, syntax, and imagery.

In the first section McCarthy uses a sequence of participles, the active "chasing" balancing the passive "watching" and "drifting," to follow the shifting contours of the rangy melody:

> I'm always chasing rainbows,
> watching clouds drifting by

McCarthy then makes a smooth lyrical transition to the next musical section by following the final *by* with a rhyme, *my,* that opens the second eight-bar sequence, then further pulls both sections together with a final rhyme on "sky" that echoes the initial "*I'*m":

> My schemes are just like all my dreams,
> ending in the sky

In the third section, McCarthy shifts to "*I'm*" rhymes, and modulates his syntax from participles to another *-ing* part of speech, the gerund:

> Some fellows look and find the sunshine,
> I always look and find the rain
> Some fellows make a winning some*time,*
> I never even make a gain, believe me

The casual conversational tag, "believe me," functions as a bridge back into the final A section, which opens with "I'm," a rhyme that connects back to the "some*time*" of the preceding section.

Not only does McCarthy work within the structure of the thirty-two-bar formula to produce a wistfully understated lament that is a far cry from "After the Ball," the character who gives voice to the lyric, the "I" of the song, is equally formulaic. This thoroughly American rainbow-chaser would become the perfect "voice" for wittily turned lyrics that balance nonchalance and sophistication, slang and elegance. (While McCarthy makes this "I" male—"some fellows look and find the sunshine"—lyricists usually tried to keep that voice androgynous, for the simple reason that a song stood a better chance of becoming popular if it could be performed and recorded by both male and female vocalists). In one guise or another, this character would sing most of the standards of the golden age, from "I'm Forever Blowing Bubbles" (1919) to "It Never Entered My Mind" (1940). As late as 1941, Tom Adair could still touch up McCarthy's 1918 proto-

type, tightening the imagery a bit here, loosening the slang a bit
there:

> I make a date for golf,
> and you can bet your life it rains,
> I try to give a party
> and the guy upstairs complains,
> I guess I'll go through life,
> just catchin' colds and missin' trains

Given the many changes we shall see lyricists ring upon this formulaic model of saying "I love you" in thirty-two bars, it seems fitting
that Adair's complaint—the "standard" complaint—should be
"Everything Happens to Me."

Ragged Meter Man:
Irving Berlin

Irving Berlin has no "place" in American music; Irving Berlin IS American music.

JEROME KERN

One day in 1909 the fledgling lyricist Irving Berlin presented his latest effort to an Alley publisher. The lyric was about "Do-rando," a famous Italian runner who had just lost the Olympic marathon, and the publisher was eager to have a hot topical song. "I'll take it," he said, then added, "You *do* have music to go with these words, don't you?" Berlin, who could barely pick out a melody with one finger on a piano, gave a hedging "Oh, of course." He was then ushered into the office of one of the staff pianists. "Hum it to him," the publisher said, "And he'll play it for you." In the time it had taken him to cross the hall, Irving Berlin had concocted a melody to fit his words. From then on, Irving Berlin the lyricist became Irving Berlin the composer—one of the few songwriters on Tin Pan Alley who wrote both words and music.

With such versatility, speed, and sheer *chutzpa*, Irving Berlin was able to attune his musical and lyrical style to every changing fashion on Tin Pan Alley, and to follow his career is to trace the history of American popular music for over half a century. When ragtime songs were in, Berlin wrote dozens of them, from "Stop That Rag" (1910) to "That International Rag" (1913). For World

War I, he wrote "Oh! How I Hate to Get Up in the Morning," then came up with "This Is the Army, Mr. Jones" for World War II. When Ziegfeld needed music for his lavish spectacles, Berlin gave him more songs than Ziegfeld had girls, including "A Pretty Girl Is Like a Melody," which became the theme song of the *Follies*. With the advent of the thirty-two-bar ballad, Berlin outdid all other Alley songwriters in saying "I love you" within the AABA formula; in the 1920s he wrote so many artfully simple romantic laments, from "All by Myself" to "How About Me," that Cole Porter dubbed the entire genre the "Berlin ballad." When the talkies came in with *The Jazz Singer,* Jolson sang Berlin's "Blue Skies" from the hitherto silent screen. For the radio crooners of the early 1930s, Berlin supplied "Say It Isn't So" (for Rudy Vallee) and "How Deep Is the Ocean" (for Crosby'). When Hollywood turned elegant with Astaire and Rogers, Berlin kept step with "Cheek to Cheek" and "Top Hat, White Tie, and Tails." Even in the 1940s, when Broadway musicals had to be "integrated," Berlin came up with songs that fitted the characters and story of *Annie Get Your Gun* almost as quickly as he'd put music to his lyric for "Dorando" forty years earlier. Not to mention, of course, his secular holiday songs—"White Christmas," "Easter Parade," and the all-purpose "Happy Holiday." Nor the unofficial national anthem, "God Bless America," which Berlin had written in 1918, filed away, then dug up in 1938, when Kate Smith asked him for a patriotic song.

It seems only fitting, therefore, that Irving Berlin should arrive in America in 1892, the year Tin Pan Alley had its first million-seller in "After the Ball." He was four years old then, and his name was Israel Baline, the youngest child of Russian Jews fleeing a pogrom in their Siberian village of Temun. The Balines settled in the tenements of the Lower East Side, where Moses Baline, who had been a cantor in Russia, now had to work in a kosher meat factory, give Hebrew lessons, and still struggle to support his family.

When his father died, eight-year-old "Izzy" quit school to sell newspapers in the Bowery. From the street, he could hear the hits of the day drift through the doors of saloons and restaurants, and he soon found that if he sang the songs himself while he sold papers, people tossed coins to him—a sign of things to come. To his mother one night he confessed his life's ambition—to become a singing waiter in a saloon. This—from the son of a cantor!

When Leah Baline objected, Israel ran away from home—to Tin Pan Alley. He broke in as a plugger, leaping up from the back rows of Tony Pastor's vaudeville house to lead the audience in refrains of the latest hits. At sixteen he finally landed the job he had dreamed of—a singing waiter at the Pelham Cafe in Chinatown. The owner, a swarthy Russian nicknamed "Nigger Mike" Salter, came to "Izzy" and another waiter one day in 1907 and told them to write a song. Two waiters at a rival cafe had just written a hit novelty song, "My Mariuccia Take a Steamboat," and, he thought, anything those waiters could do, his could do better.

Berlin wrote a lyric chock full of the worst clichés,

> My heart just yearns for you!

tortured inversions,

> Please come out and I shall happy be!

and strained poeticisms:

> 'neath the window I'm waiting

Still, there were touches, here and there, of the real language of the streets:

> Please don't be so aggravating

The song was called "Marie from Sunny Italy," and the royalties from the sheet music sales netted Israel Baline thirty-seven cents. But the sheet music brought an unexpected bonus: the name Irving Berlin. The typesetter, so the story goes, mistakenly had listed the lyricist as I. Berlin but Izzy Baline found the misprint fortuitous: it had *class*. To give it even more class, he made the "I" stand for Irving.

He had been picking out tunes on the cafe piano after closing (he never would learn to play on anything but the black keys,
though soon he got a piano with a special lever that allowed him
to transpose into other keys). A year later, after the success of
"Dorando," he began writing his own music. He was fascinated
by ragtime—even as its demise was being touted—and quickly
turned out a series of hit rags, from "Yiddle on Your Fiddle, Play
Some Ragtime," through "That Opera Rag," to "Oh, That Beautiful Rag." It was "Alexander's Ragtime Band" (1911), however,
the biggest hit Tin Pan Alley had yet seen, that gave ragtime new
life and brought Berlin international fame.

It has been pointed out countless times that "Alexander's
Ragtime Band" is *not* really ragtime, since it contains barely a
hint of syncopation, but no one has noted how skillfully "ragged"
the lyric is. What ragtime did for song lyrics generally during the
early years of Tin Pan Alley it did for Berlin as well. It licensed
the vernacular as a lyrical idiom and forced the lyricist to construct a lyric out of short, juxtaposed phrases marked by internal
rhymes and jagged syntactical breaks:

> Ain't you goin',
> ain't you goin',
> to the leader man,
> ragged meter man?
> Grand stand,
> brass band,
> ain't you comin' along?

The "ragged meter man" here is not only Alexander—a code
name, going back to "Alexander, Don't You Love Your Baby No
More" (1904), which stamps this as a "coon" song—but Berlin
himself, who fits slang contractions and telegraphic exclamations
to the uneven, bugle-call snatches of his musical meters. Just as
he juxtaposed off-rhymes like *leader* and *meter*, Berlin ragged the
normal accent of "natural" for an "unnatural" rhyme with "call":

> They can play a bugle *call*
> like you never heard before,

So natur*al*
that you want to go to war

Years later, in the 1960s, Berlin would change this last line to the less martial "so natural that you want to hear some more," but he still kept ragging his rhymes with *natur*al and th*at you*.

Because he wrote both words and music, Berlin could keep repeating, yet subtly varying, a series of short, parallel phrases—both musical and lyrical phrases—to build a sense of urgency and excitement:

Come on along, come on along,
let me take you by the hand,
up to the man, up to the man

The image, like the language, here evokes a revival camp meeting, but we never do get to meet the maker of the music. At the end of the chorus Berlin deftly ducks away from climax with a long subordinate clause—"if you care to hear the 'Swanee River' played in ragtime"—but that sonorous lyrical line is brassily interrupted by a syncopated stop before "played." The final "Come on and hear" thus comes off as casual invitation rather than fervid plea.

The international success of "Alexander's Ragtime Band" gave ragtime new life and sparked a national dance craze. The leaders of that craze were Irene and Vernon Castle, and in 1914 Berlin wrote a revue in ragtime, *Watch Your Step,* which showcased their talents. Although he had placed songs in other Broadway productions, *Watch Your Step* was his first complete score, and its songs radiated the musical and lyrical sophistication Alley wares acquired when they emanated from the theater. Ragtime had quickly come to signify modernism, and Berlin caught the cultural struggle between Victorian gentility and the purveyors of liberation, indulgence, and leisure in "Play a Simple Melody." This was the first of his famous "double" songs in which two different melodies and lyrics are counterpointed against one another. The first melody is fitted to a nostalgic call for a song "like my mother sang to me," but that genteel request is juxtaposed against a brassily vernacular demand:

> Won't you play me some rag?
> Just change that classical nag
> to some sweet beautiful drag

Such an intricately woven clash of music and lyric raised the principle of ragging to a larger structural level.

Ironically, in the year between "Alexander's Ragtime Band" and "Play a Simple Melody," Berlin had experimented with the very kind of sentimental ballad that ragtime had displaced. After the sudden death of his wife in 1912, only five months after their marriage, Berlin poured out his grief in "When I Lost You." The song sold well, yet its lyrics are nearly as poetically strained as "Ah! Sweet Mystery of Life":

> The sunshine had fled,
> the roses were dead,
> Sweetheart, when I lost you

Clearly Berlin still saw the slang of ragtime as an inappropriate idiom for serious romantic expression; not until the 1920s would he adapt its colloquial idiom for love songs.

How much fresher the language of his comical ragtime love song of 1915, "I Love a Piano," with its clever ragging of words against music, such as crushing "pian" into one syllable for "upon a *pian*-o," then drawing it out properly for "a grand pi-a-no," giving back to "grand" some of its original currency. In this song, too, Berlin, almost as if he had been reading the new "imagist" poetry that had begun appearing in *Poetry* magazine, develops an extended metaphor that implicitly compares the piano to a woman. The cliché "a fine way to treat a lady" turns into "I know a fine way to treat a Steinway," and the various parts of the instrument are transformed into erogenous zones by the singer's exuberant passion:

> and with the pedal
> I love to meddle

or

> I love to run my fingers o'er the keys,
> the ivories

Here the poetic "o'er" is quickly redeemed—and rhymed—by its slangy gloss, the "ivories." At the climax of the song the very letters that name the beloved object elicit a lingering ejaculation:

> Give me a P-I-A-N-O, Oh, Oh

Behind the piano-as-girl metaphor Berlin again counterpoints the genteel and the modern sensibilities as the "upright" and "high-toned" piano is exuberantly and irreverently fondled as a "baby" grand.

In 1918 such witty imagery enabled Private Irving Berlin, in the army now, to invoke the slang he must have heard in his barracks at Camp Upton, New York. The hit of his all-soldier revue, *Yip, Yip, Yaphank,* was "Oh! How I Hate to Get Up in the Morning," which, along with "Over There," was one of the few patriotic Alley songs to survive World War I. In place of Cohan's thumping jingoism, Berlin substitutes the ordinary soldier's perspective on the drudgery of army life. He develops the song from the helpless lament of the title through other slang formulas like "ya gotta get up" and the vernacular threat:

> Someday I'm going to murder the bugler;
> someday they're going to find him dead

Berlin then uses a bizarre image to suggest, however tangentially, the sort of threat a real soldier would make:

> I'll amputate his reveille
> and step upon it heavily

The Latinate "amputate" softens the soldier's pithier "I'll cut off his . . . ," and the French "reveille" merely hints at the instrument the soldier might threaten to sever in pure, but unprintable, Anglo-Saxon.

Berlin wrote hundreds of songs before 1918, but most were topical songs that enjoyed only brief popularity and have not survived as standards. Some were for the new dances, from the grizzly bear to the chicken walk and the fox trot. When a dancer named Doraldina started a Hawaiian craze in 1916, Berlin came up with "That Hula-Hula," and, since the craze started by Little Egypt at the Chicago Exposition of 1893 was still going strong in 1915, he followed step with "Araby." He did his obligatory string of southern songs—"When the Midnight Choo-Choo Leaves for Alabam'," "When It's Night Time in Dixie Land," "When You're Down in Louisville"; pastorals—"I Want to Go Back to Michigan—Down on the Farm," "This Is the Life," and "Si's Been Drinking Cider"; travelogues—"San Francisco Bound" and "From Here to Shanghai"; immigrant songs—"It Takes an Irishman to Love" and "I'll Take You Back to Italy"; and, of course, more rags—"Whistling Rag," "Ragtime Violin," "Ragtime Jockey Man," "Ragtime Soldier Man," so many that when he entertained fellow passengers on a transatlantic crossing they asked him how he knew so many ragtime hits ("I wrote them," the always-modest Berlin admitted).

Amid all of this staggering output, however, one finds few love songs; when love is the subject, it is usually handled comically, as in "If That's You're Idea of a Wonderful Time, Take Me Home" or "Keep Away From the Fellow Who Owns an Automobile." Berlin had broken into the Alley just as the ballad was being revived, but it was not until after the First World War that he devoted himself to fashioning romantic laments, pleas, and effusions to the by-then standard framework of the thirty-two-bar AABA chorus.

A key song in this transition from rags and topical songs to lyrical ballads was one that Berlin wrote for Ziegfeld's *Follies of 1919*, "A Pretty Girl Is Like a Melody." Calling it one of Berlin's "first big guns," Alec Wilder puts it "on a level with Kern's pure melodies" and, comparing it with the first decade of Berlin's music, finds it "extraordinary that such a development in style and sophistication should have taken place in a single year."[2] The lyric is equally innovative, elaborating a single, extended simile in casually colloquial terms: a pretty girl is like an insistent melody that at first "haunts," then starts a ghostly "marathon" that gives you the "runaround," and finally produces both musical and athletic "strain," as her fleeting image reverses itself and imprisons its pursuer. At the end, you "can't escape" because, paradoxically, "she's in your memory." Berlin's "pretty girl" is not so much flesh-and-blood but a tantalizing mental image:

> She will leave you
> and then
> come back
> again

By breaking up the lyrical phrase to match the musical pauses, Berlin captures the elusive play of fantasy; little wonder that the song became the theme for Ziegfeld's revues, some of which supplied patrons with balloons for playing "catch" with the coyly elusive chorus girls or lariats with which to "rope" them.

For revues at his Music Box theater, which opened in 1921 to packed houses despite the exorbitant ticket price of five dollars, Berlin composed an equally lush song, "Say It With Music." Giving yet another twist to the metaphoric equation between romance and music, Berlin took an advertising formula and elaborated it with nonchalant turns of American talk

> Somehow they'd rather be kissed
> to the strains of Chopin or Liszt

Even though he had abandoned ragtime songs, Berlin found he could still adapt the technique of ragging words against music. Using parallel whole notes to stretch out the first syllables "beautiful" and "music," for example, gave him a subtle but sonorous rhyme. The triple "el" rhyme that runs through "melody mellow played on a cello" sets up a quiet fourth rhyme when "Helps" is drawn out by another whole note.

Such distortions were the lyrical equivalents of the "syncopation" that Berlin theorized freely about to reporters and columnists in the 1920s. Syncopation, he affirmed, was the characteristic style of modern American music; its "broken harmonies" and "ragged time" set a "new rhythm," a "new method of movement" that brought modern music in tune with the age of the "automobile."[3] Much of the lyrical artfulness of Berlin's "sob-ballads," as he called them, stems from his subtle fragmentation and juxtaposition of words *against* music.

The newspapers, however, interpreted such hits as "All Alone" (1924) and "Remember" (1925) in the light of Berlin's courtship of socialite Ellin Mackay. Berlin insisted, however, that the only song he ever wrote out of personal experience was "When I Lost You"; the "sob ballads" were merely his new line of wares for a public that now, in the 1920s, "would rather buy tears than smiles."

While he continued to write novelty and topical songs for revues in the '20s, his popular hits—those written strictly for sheet music sales—were almost all romantic laments. Many were waltzes, such as the mawkish "Always" (1925), which Berlin gave to Ellin Mackay as a wedding present (but which George S. Kaufman cynically suggested be retitled "I'll Be Loving You—Thursday"). In the best of these ballads, however, there is an easy "syncopation" of vernacular phrases. One of the earliest of them, "All by Myself" (1921), takes the simple catch-phrase of the title and lifts it out of its ordinary context as a child's boast ("I did it all by myself") into a literal scene, as barren and simplified as the language that describes it:

> I sit alone
> with a ta-
> ble and a chair
> so un-
> happy there
> play-
> ing solitaire

By breaking up the verbal phrases in this way, Berlin's "ragging" produces fragments that rhyme in unusual ways: the first sylla-

bles of "t*a*ble" and "pl*a*ying," then an off-rhyme between the
second syllable of "al*one*" and "*so un*happy."

Berlin was reworking the traditional figure of the forlorn
lover, placing him in a room as empty as a prison cell, and, three
year later, with "All Alone," he injected some humor into the
uncluttered scene:

> waiting
> for a ring,
> a ting,
> a ling

The repetitive rhymes capture the obsessive sensibility of such a
prisoner of love, and Berlin's insistent folding of sound fragment
around sound fragment tightens the psychological chains. Even
the catch-phrase title, "*all al*one," is a ready-made instance of
faceted, repetitive sound fragments. Still another raggedly repetitive pattern closes the song, as Berlin at first matches three musically parallel phrases with two lyrically parallel ones:

> wond'ring where you are,
> and how you are,

But then, as Gerald Mast observes, the third lyrical phrase is a
"syntactic surprise,"[4] which leaves the thought dangling,

> and if you are

only to be completed by the standard return to the title: "all
alone too."

Berlin's other big hit from 1924, "What'll I Do?," also evokes
a solitary singer, alone in his room, obsessively lamenting his fate
in repetitive colloquial phrases. In this song Berlin uses conversational contractions—"What'll" and "wond'ring"—yet alternates
them with formal spelling-outs of the more common contractions "I'm" and "who's" to create a jagged pattern of rhymes:

> What'll I *do*
> when I

am won-
d'r*ing*
wh*o*
is
k*iss*-
ing
y*ou,*
what'll I *do?*

By contracting "wond'ring" Berlin creates a rhyme on the "ing" of "kissing," while "kiss" rhymes with "is" only because he spells out the expected contraction "who's" into "who is."

Then he switches back to contractions with "I'm," so he can fit a triplet to the phrase,

When-I'm-a-

thus isolating

lone

over a half and an eighth-note. A bar later he uses the half-note–eighth-note pattern to split

on-
ly

so that the last syllable of "*alone*" rhymes with the first of "*only.*"

In another phrase, "what'll I do with just a photograph to tell my troubles to," he rhymes homonyms but underlines their difference musically by placing one "to" over a short eighth note, the second over long half-notes. A musical tie also binds that first "to" to the one embedded in

pho-
to-
graph

which also takes an eighth-note. That embedded "to" in "photo-graph," however, only forms an eye-rhyme with the other two "to's"; Berlin places its true rhyme in the preceding measure:

> with just
>
> a

He marks that rhyme musically by placing "a" over the same eighth-note as the "to" of "photograph."

Significantly, such intricately worded ballads as "What'll I Do?" and "All Alone" were first heard not on the stage but over the radio, sung by Frances Alda in a broadcast honoring Irving Berlin in 1924. Not only were they introduced over a new medium, they were among the first songs whose record sales equaled their sheet music sales, marking a shift in the very nature of Tin Pan Alley—from a row of music publishing houses to a network of radio, re-cording, and, soon, Hollywood sound studios. Not only did the ra-dio and the phonograph signal a new medium for popular songs, they invited, as Berlin seems to have realized, a new kind of song as well, a more intimate one that responded to the trend of Ameri-can mass culture in the 1920s, as analyzed by Lewis Erenberg, toward a preoccupation with private experience.[5] Traditional bal-ladeers sang stories to audiences, and the early Alley songs were similarly geared for either the stage audience or for the group sing around the parlor piano piled high with sheet music. Berlin's ballads of the '20s, however, imply a solitary listener, at the phono-graph or radio, and his technique of folding the tiniest rhyming fragments over and over one another creates a lyrical "space"—self-enclosed, repetitive, faceted—that is designed for the self-absorbed, plaintive singer who inhabits it. The solitary consumer of "sob ballads," in turn, inhabits the same space, the space that T. S. Eliot described so bleakly in *The Waste Land:*

> When lovely woman stoops to folly and
> Paces about her room again, alone,
> She smoothes her hair with automatic hand,
> And puts a record on the gramophone.

The bare room, right down to the gramophone, is the same one
sketched in Berlin's ballads; the song Eliot's typist listens to could
easily be "All by Myself."

As simple as the ballad form was, Berlin, because he wrote
both words and music, could treat it with more intricacy and
subtlety than most other Alley lyricists of the 1920s. In "Blue
Skies" (1927), for example, he plays with the ambivalent connotations of "blue." In the verse the singer recalls,

> I was blue, just as blue as I could be;
> ev'ry day was a cloudy day for me

but in the chorus she exults,

> Blue skies! Smiling at me!

with the characteristically ragged rhyme between "sk*i*es" and
"sm*i*ling." Musically, however, Berlin cuts against the stated happiness of the lyric by beginning each A section in the minor key,
then staying in it for five of the eight measures, so that the song
sounds sad and "blue" even as its words celebrate gloriously
"blue" skies.

This musical counterpoint to the lyric is subtle, but the lyric is
equally deft at undercutting the very happiness it affirms. It does
this, in part, by the preponderance of negative terms—"Nothing
but blue skies," "nothing but bluebirds," "never saw the sun,"
"never saw things"—even the neutral "noticing" seems negative in
this context. Conversely, the repeated pronoun "nothing" is
drawn by its rhyming suffix into the constellation of participles—
"smiling," "singing," "shining," "going"—that at first seem to give
duration to the happy present, but then, in the release, only underscore its transience:

> Noticing the days, hurrying by,
> when you're in love, my! how they fly

The melancholy cliché that "to be conscious of present happiness
is to be conscious as well of its mutability" is the Keatsian pivot
that turns blue into "blue." Yet Berlin couches this minor-key

pall in nonchalant slang: "my! how they fly." In the final A section, he can dangle an equally colloquial formula—"Blue days—
all of them gone" to suggest both loss and relief.

In some of his streamlined ballads, such as "How About Me"
(1928), Berlin even dispensed with a verse, "surprisingly early,"
notes Wilder, since "the fashion for not writing verses did not
begin until much later."[6] The static pathos of the helpless singer
is barely alleviated by the aggressive slang formula of the title.
Another slang formula—the Yiddish penchant for answering a
question with another question—redeems the near-monotony of
"How Deep Is the Ocean" (1932), by "answering" such inquiries
as "how much do I love you?" with such exasperated replies as
"how high is the sky?" Berin took the title question out of an
earlier lyric, "To My Mammy," then followed Elizabeth Barrett
Browning's lead in "How Do I Love Thee" by spinning a series of
questions into the teasing symmetry of a children's riddle, closing
with the opening queries:

> And if I ever lost you,
> how much would I cry?
> How deep is the ocean?
> How high is the sky?

Just as the ballad lent itself to the medium of phonograph
records in the 1920s, it was equally effective for the radio crooners of the 1930s. When Rudy Vallee introduced "Say It Isn't So"
(1932) on his radio show, the song not only became an overnight
hit, it saved Vallee's marriage. (The Vallees had planned to get a
divorce, but after Vallee sang Berlin's lyric, both he and his wife
dissolved in tears and called off the "calling-off.") Berlin feeds
the crooning style with long, lingering notes fitted to expansive
open vowels, like the *a* and *o* sounds of the title phrase. Still, he
manages to avoid monotony by containing the repetitive musical
phrases in an unusual ABCD structure; lyrically he frames the
insistent *a* and *o* oscillations in utterly colloquial formulas, such
as "Say that everything is still okay" and "That's all I want to

know." In the final lines of the chorus he creates an even slangier feel to the lyric by dropping syntactic connections:

> And what they're saying—
> say it isn't so

Thus the lyric itself "doesn't say" what it longs to have "unsaid."

Although ballads dominated Berlin's work as much as they did the output of Tin Pan Alley generally, he continued to write upbeat songs for revues. In these songs he toyed with the same rhythmic disruptions of lyrical phrases that marked his ragtime songs. In "Everybody Step" (1921), for example, he used a sequence of three sets of four eighth-notes each to split the lyric into four fragments that create rhyme by rhythmic accent:

> Syncopated
> *rhythm!*
> Let's be goin'
> *with'em!*

Similarly, he enlivened the rather melodramatic apostrophe to a "Lady of the Evening" with an extended simile reminiscent of Longfellow but made more striking by its rhythmic divisions:

> You can make the
> cares and troubles that
> followed me through the
> day
> fold their tents just
> like the Arabs and
> silently steal a-
> way

Here the shifting musical and lyrical phrasing mimes the elusive image it conjures up. Similarly, in "Shaking the Blues Away" (1927), he used rhythmic accents to mark such contractual rhymes as "you'll" with "rule" and—like Rourke's lyric for "They Didn't Believe Me"—"they'll" with "tell."

Berlin's greatest rhythmic song of all, however, was written
not for a Broadway revue but a Hollywood movie, *Puttin' on the
Ritz,* in 1930. The title song carries the principle of lyrical ragging to the furthest possible extreme, so distorting verbal accents
against musical ones that the lyric, when sung, comes out as
Gertrude Steinese or a jazzy idiom that might be called Berlintz:

> *Come* let's *mix*
> where *Rock-e-fell-*
> ers *walk* with *sticks*
> or *um*-ber-*el-*
> las *in* their *mitts,*
> *put*-tin on the *ritz!*

Here the musical accents break down sentence, phrase, and word
into tiny Cubistic fragments fitted "mosaically" to musical shards.
The sharp rhymes only highlight the discordant levels of diction—
"ritz" and "Rockefellers" clashing with "mitts" and "puttin," and
the elegant walking stick reduced to the prosaic "walk with sticks,"
while the ordinary umbrella is elevated by enunciated elongation.

Such a lyric realizes the possibilities of "ragging" that opened
at the turn of the century with such "coon" songs as "Under the
Bamboo Tree," and Berlin must have had those origins in mind,
for, in one of the sets of lyrics he provided for "Puttin' on the Ritz,"
he celebrated not the "well-to-do" on "Park Avenue" but blacks
parading on "Lennox Avenue" where "Harlem sits":

> *Spang*led *gowns*
> up*on* a *be*vy
> *of* high *browns*
> up*on* the *le*vee—
> *all* mis-*fits,*
> *put*tin on the *ritz!*

What the lyric describes—Harlem blacks in elegant finery—it
also enacts in its own linguistic "mix" of slang and refined diction
and allusion, its clever "misfits" of rhythmic and verbal accents.

It is ironic that "Puttin' on the Ritz," one of Berlin's best lyrics, was written during a period when he began to fear his creative powers were waning. The very titles of his recent—and perhaps, he thought, last—hits, "The Song Is Ended" and "Where Is the Song of Songs for Me?" seemed to express the fear that he could write no more. Although Berlin's dejection was largely personal, it is significant that it should come at another turning point in the history of popular music. The simple, repetitive musical style of the 1920s was giving way to more sophisticated adaptations of jazz and blues styles by composers like Gershwin and Arlen. The ante was upped for lyrics too. In the hands of Lorenz Hart and Ira Gershwin, songs began to radiate sophisticated sentiments and witty rhymes. It is significant that when Berlin found himself unable to meet his commitment to write songs for the show *Fifty Million Frenchmen* in 1929, the task went to a relative newcomer, Cole Porter, whose lyrics, even more than Hart's or Gershwin's, prefigured the insouciance, the urbanity, the antiromantic stance of the height of the golden age.

A number of lyricists who flourished during the 1920s saw their seemingly endless supply of hit songs dry up after 1928; and for a time, it seemed, Irving Berlin would fade away as well. He had already begun to sound like an elder statesman, railing against "swing" music and "sophisticated" lyrics, yet in 1933 he made a spectacular comeback on Broadway with *As Thousands Cheer*, the longest running show of the year. Based on sections of a daily newspaper, it featured Ethel Waters in the "weather report" doing a saucy—and swinging—"Heat Wave" with such uncharacteristically naughty lines as "she started the heat wave by letting her seat wave" that give a Porterish overtone to the "heat" of the title. In the "news" section Berlin gave Waters "Supper Time," a Southern black woman's lament for her lynched husband, a song that showed Berlin, too, like Gershwin and Arlen, could adapt the idiom of the blues into popular song. The big hit of the show, however, was the "fashion report" song, a throwback to the good old days. The music for "Easter Parade" had been written in 1917 for a song called "Smile and Show Your Dimple," but it was a flop then, and Berlin had judiciously filed it away. In the 1930s, he updated it with a new lyric, one that evokes a bygone era with "the quaint image of the rotogravure," which, as Timothy Scheurer notes, gives the song "a feeling of being bathed in sepia tones."[7]

Berlin stayed on Broadway only long enough to establish himself as definitely *back;* for most of the decade, he wrote in Hollywood, which by then was rivaling both Broadway and the radio as an outlet for popular song. While Broadway songs were becoming increasingly sophisticated, the movies wanted simpler fare and offered him the perfect opportunity to ply his artful artlessness by writing songs for the films of Fred Astaire. On the one hand, melodies for Astaire had to be kept simple—his voice had barely more than an octave's range—yet his dancing required rhythmic intricacy and his suave but colloquial character required casual but urbane lyrics.

In his discussion of Berlin's music, Alec Wilder observes that all songwriters were

> vitalized by Astaire and wrote in a manner they had never quite written in before: he brought out in them something a little better than their best—a little more subtlety, flair, sophistication, wit, and style, qualities he himself possesses in generous measure.[8]

Wilder cites "Cheek to Cheek" as "a case in point," but such a song instead seems to reveal how even Berlin sometimes found it difficult to come up to Astaire's standard lyrically as well as musically. While the musical case certainly stands (Berlin's melody has an unusual length—72 measures—and structure—AABCA with no verse and subtle shifts from major to minor), the lyric presents a series of trite images of male sportsmanship set against the "effeminate" activity of dancing:

> Oh! I love to climb a mountain
> and to reach the highest peak . . .
> Oh! I love to go out fishing
> in a river or a creek

Equally banal are the clichés "my heart beats so that I can hardly speak" and "I seem to find the happiness I seek." To be sure, the metaphor of cares disappearing "like a gambler's lucky streak" is a little better, and the opening, "Heaven—I'm in heaven," gives a slangy abruptness to another of Berlin's verseless songs.

Significantly, it is in the C section, which Wilder finds musically "amazing," that Berlin falls back lyrically upon the tiredest of rhymes, the "charms about you" and "my arms about you" barely redeemed by the pun on *about*. Nevertheless, his talent for juxtaposing lyrical against musical phrasing comes out as he moves back to the final A section. The "charms" carry both the lover and the release "through to" the opening phrase of the last section, "Heaven, I'm in heaven," with the "to" not only bridging the gap syntactically but rhyming with "through."

A much better case for Berlin's ability to match Astaire's style would be the title song for the same film, *Top Hat*. "Top Hat, White Tie, and Tails" is another extreme instance of musical accents ragging verbal ones, though the distortions are not so intricate as they are in "Puttin' on the Ritz." They mostly occur in the verse, where the misplaced accents give a lilt to the words that reflects the buoyancy of the singer:

> *I* just *got* an *in*vita-
> tion *through* the *mails*

then the staggered, fragmented lines of the formal invitation itself:

> your *pres*ence
> re*ques*ted
> this *eve*ning
> it's *for*mal

The chorus continues to fragment words against music but lingers over initial syllables of participles to bring out the *i*-rhymes: "Ty-in' up my white tie."

In the release yet another rhythm rags the long formal phrase into short units.

> I'm steppin' *out*
> my *dear*
> to breathe an *at-*
> mos*phere*

which then are followed by an abrupt shift to street slang:

> that simply *reeks* with *class*

Berlin clashes the same high and low diction in the next pair of phrases, the elegant "I trust that you'll" jammed against the slangy "excuse my dust when I step on the gas." As in "Puttin' on the Ritz," the rhythmic distortion of formal diction is reflected in the subject of the song: at the end, the singer gleefully looks forward to "Puttin' down my top hat" and "Mussin' up my white tie." The final line, "Dancin' in my tails" is a good description of how Berlin has made the formal language dance against the music and the jostling intrusions of street slang.

In a different but still thoroughly Astaire vein was "Change Partners," which lent itself perfectly to his seductive nonchalance. "*Must* you dance—*every* dance—with the same—fortunate man," Astaire asks in elegant exasperation, exasperation underscored by buried rhymes in d*a*nce and m*a*n. Berlin also reaches back to his 1920s ballads for similar rhymes and imagery in the release, but here they serve Astaire's coy sophistication:

> Ask him to sit this one out
> and while you're alone,
> I'll *tell* the waiter to *tell* him
> he's wanted on the *tel*ephone

Equally urbane is Astaire's understated final plea—"change partners and then—you may never want to change partners again."

Berlin continued to write film songs—and hits—throughout the 1930s and found Hollywood a haven where, as one of his songs put it, "An Old-Fashioned Tune Always Is New" (1939). In 1940, however, he returned to Broadway with *Louisiana Purchase,* a show that, even though it produced no major hits, looked forward, in its regional emphasis, to *Oklahoma!.* Returning to the theater must have made Berlin feel a little like Rip Van Winkle: just as Rip had slept through the American Revolution, Berlin had waited out the golden age of Manhattan urbanity in the California sunshine. Where Rip returned to find everything utterly changed, however, Berlin found the new style in popular songs much as it had been twenty years earlier. Topical songs had returned with the war, and Berlin's first popular songs were as made to order as they had been in World War I: "A Little Old Church in England" and "Any Bonds Today?" Back again were songs about the South, and other places, like Vermont and Capistrano, Kalamazoo and Kansas City. Once more there were lyrics about trains and airplanes, cowboys and soldiers; even birds were back—skylarks, bluebirds, swallows—birds that hadn't been seen in the Alley for over a decade.

Back, too, was simplicity and sincerity: a sentimental song of 1931, "As Time Goes By," was revived from oblivion and became a tear-jerking hit in *Casablanca.* Going, if not gone, were insouciance, cynicism, and sophistication, along with some of their purveyors, such as Lorenz Hart, dead in 1943—the man who sniffed in congealing love the "faint aroma of performing seals"—replaced by a cock-eyed optimist who even loved the whiskers on kittens. Berlin himself pronounced the era's epitaph, noting that nothing is "so corny as last year's sophistication," adding, "I mean corny lyrics . . . there's no such thing as a corny tune."[9] A decade that had begun with Ira Gershwin's flippant post-Crash patriotism in "Of Thee I Sing—Baby" (1931) closed with Berlin's solemn look at impending war in "God Bless America" (1938). Berlin had his big Broadway success in 1942 when he reincarnated *Yip Yip Yaphank* as *This Is the Army,* replete with new-old songs, and, as if to symbolize the unchanging changes, Berlin appeared in his old World War I uniform (which still fit) to sing "Oh! How I Hate to Get Up in the Morning." In the same year he wrote "White Christmas" for Hollywood, tugging again at the nostalgia he had milked in "Easter Parade."

In one way, however, popular songs had changed profoundly, and Berlin once more showed his ability to change with the changing times. With the 1943 production of *Oklahoma!*, Broadway songs assumed a new character that they have maintained ever since. No more the "interpolated" song of '20s and '30s musicals that had so little relevance to the plot or characters of the libretto it could be freely shunted from one show to another; theater songs after *Oklahoma!* had to be "integrated" into character, dramatic context, and sometimes even had to advance the plot.

That Irving Berlin, nearly sixty, was able to write any theater songs in the new style is impressive; that he was able to write one of the greatest "integrated" musicals is remarkable; that he was able to write it in the space of a few weeks is astounding. *Annie Get Your Gun* was produced by Rodgers and Hammerstein in 1946. They had commissioned Dorothy and Herbert Fields to write the book and Jerome Kern to do the music. When Kern died suddenly, they turned to Berlin. According to Martin Gottfried, Berlin

> didn't quite believe the producers' excuse that they were "too busy with another project" to write this one themselves. Rodgers, he thought, can write anything, so Berlin concluded that Hammerstein considered *Annie Get Your Gun* too superficial an entertainment for the team. Berlin was also uncertain that he could write lyrics for the rural characters in *Annie Get Your Gun*. Hammerstein assured him, "All you have to do is drop the 'g's.' Instead of 'thinking,' write 'thinkin'.' " Berlin gave it a try, going home and writing "Doin' What Comes Natur'lly." Though Rodgers and Hammerstein approved, he still wasn't certain, went home again, and wrote "They Say It's Wonderful."[10]

In the next few hectic weeks he came up with a string of superb songs—"Anything You Can Do," "You Can't Get a Man With a Gun," "I Got the Sun in the Morning," "I Got Lost in His Arms."

The most famous song from *Annie Get Your Gun,* "There's No Business Like Show Business," was nearly dropped from the show. When Berlin played it for Rodgers and Hammerstein, the two men just sat there, too awed at first to speak. Mistaking their silence for disapproval, Berlin simply tossed the song aside and promised to come up with something better.

Berlin's songs from *Annie Get Your Gun* looked both backward and forward in American musical theater history. On the one hand, they nearly all became independently popular (*Annie Get Your Gun,* affirms Gerald Mast, "contained more individual hit songs than any musical ever, before or since"[11]), a throwback to the days when a musical had a book that merely served as a clothesline for Tin Pan Alley songs. But those songs also were "integrated" in the style of the new musical play inaugurated by *Oklahoma!:* they expressed dramatic situations, rendered conflict, and delineated character.

In writing such integrated lyrics Berlin displays a whole new range of writing techniques with apparent effortlessness. He can open a song quietly, with a deadpan irony whose colloquial ease belies its naughty joke: "Oh, my mother was frightened by a shotgun, they say; that's why I'm such a wonderful shot." In the space of a few bars he can build short phrases to an outraged climax:

> If I went to battle
> with someone's herd of cattle,
> you'd have steak when the job was done,
> but if I shot the herder,
> they'd holler bloody murder,
> and you can't shoot a male
> in the tail,
> like a quail,

then suddenly twist it off with the helpless rage of "Oh, you can't get a man with a gun."

The same Annie who sings that aggressive lament still guards her feelings in a love song with "They say," the same phrase she used to recollect her mother's shotgun wedding, turned now to a protective hedge:

> They say that falling in love is wonderful,
> it's wonderful (so they say)

Here, too, he recasts the "wonderful" from the braggadocio of "I'm such a wonderful shot" to underscore hesitant innocence.

Berlin clearly had been listening during the era of wit and sophistication. There is a touch of Ira Gershwin's laconically ebullient "s'wonderful" here, along with a dose of Cole Porter's colloquial understatement. In the release Berlin is pure Hart:

> I can't recall who said it,
> I know I never read it,
> I only know they tell me that love is grand,
> AND . . .

Hart buckled a musical bridge the same way a dozen years before in "Thou Swell":

> just a plot of—
> not a lot of—
> land,
> AND

But where Hart's lyric displays his own verbal pyrotechnics, Berlin's calls attention, not to itself, but to a character caught between hesitancy and eagerness.

Berlin went on to write other musicals, such as *Call Me Madam,* and continued to turn out hits in record time. He loved to write for the brassily elegant Ethel Merman ("she makes sure you can hear my lyrics in the back row of the balcony"), and when Merman suggested that the second act needed a little more punch, Berlin stayed up all night and turned out yet another of his complex double-songs, "You're Just in Love." Against the sweetly romantic strains of "I hear singing and there's no one there," Merman's chorus bristles with current jargon couched in street slang:

> you don't need analyzing,
> it is not so surprising . . .
> there is nothing you can take
> to relieve that pleasant ache;
> you're not sick—you're just in love!

and spiced with elegantly erotic touches:

> Put your head on my shoulder,
> you need someone who's older,
> a rub-down with a velvet glove

Using the musical emphasis to reverse the verbal accent here (not "*rub*down" but "rub*down*"), Irving Berlin was still the ragged meter man of Tin Pan Alley.

Berlin eased into retirement only as Tin Pan Alley itself was disappearing under the onslaught of a new kind of music that shifted the capital—and the style—of popular music out of New York. His death in 1989, at the age of 101, marked a century since enterprising sheet music publishers began to set up shop on 28th Street. Always the purely popular songwriter, Berlin could also rival the great theatrical lyricists; while Hart, Gershwin, and Porter were unexcelled in their individual stylistic range, Berlin's long and versatile career proves that anything they could do, he could do too.

4

Ragged and Funny: Lyricists of the 1920s

*It may sound a little immodest, but you'd be amazed—
perhaps you shouldn't be—that most of my songs that sold
were written in less than fifteen minutes.*

IRVING CAESAR

One night in 1924, so the story goes, composer Vincent Youmans
came up with a melody so enthralling he got his lyricist, Irving
Caesar, out of bed and begged him to put words to it. To placate
Youmans (and to get back to sleep) Caesar quickly tossed off a
"dummy lyric," promising to write the real one in the morning.
But the next morning Caesar and Youmans looked at the dummy
lyric again and decided to keep it, even though the title phrase
was never repeated—a clear violation of the Alley's axiom that "a
good lyric" was "one that states the title promptly and then *keeps*
stating it so that the public will remember it when shopping for
records and sheet music."[1]

Yet "Tea for Two" turned out to be one of the most popular
songs of the 1920s, one that typifies the musical style of that
decade, when melodies consisted of short, repeated musical
phrases—or parallel phrases repeated, step-wise, a few intervals
apart. The phrases themselves, moreover, often contain repeated
notes. Noting the "bone simple" character of 1920s music, Alec
Wilder marvels that it produced so many good songs.[2] It is
equally surprising that lyricists like Irving Caesar could set such

music so skillfully. Yet Caesar, along with Gus Kahn, Buddy DeSylva, and numerous other Alley "wordsmiths" of that decade fashioned lyrics to such rickey-tick fare as fast as "tunesmiths" could crank it out. Supplying songs for vaudeville, Broadway, or the straight sheet music and record markets, they epitomized the heyday of Tin Pan Alley, when a lyricist like Caesar could say "I love you" in thirty-two bars of AABA—and say it quickly— whether he was working with music composed by Sigmund Romberg or Jimmy Durante.

In the late 1920s, as musical styles began to change, many of these lyricists saw their astounding ability to crank out hit songs suddenly taper off. By that time, too, Broadway, with the emer- gence of Rodgers and Hart, the Gershwins, and Cole Porter, was demanding more wit and sophistication in its lyrics, while Holly- wood's new sound films, by contrast, required lyrics that were even more simple than the simplest Alley fare. Still, though he never again had a song that equaled the popularity of "Tea for Two," Caesar, like most of these lyricists, continued writing songs; indeed, at over 90 years old, he still works at his office on Seventh Avenue.

In setting the simple melodies of the 1920s, Caesar and other Alley craftsmen made shrewd use of musical repetition to portray lovers who were nervously addicted to romance, trapped on treadmills of fate, or delightfully caught up in a round of ecstasy. The music for "Tea for Two," for example, is as symmetri- cal as the letter T itself, and Caesar opens the chorus with a dovetailed dove tale of newlywed bliss:

> Picture you
> upon my knee;
> just tea for two
> and two for tea

The musical pattern, which Wilder describes as "two-measure imitations,"[3] begets a lyric of mirror-like reversals of phrasing and see-saw oscillations between *ee* and *oo* rhymes. But just when the song threatens to babble off into honeymooner baby-talk, the music suddenly climbs up six intervals between the two syllables of "a-lone," that normally mournful term, so stretched, sounding surprisingly exuberant.

An even more powerful antidote to musical (and marital) monotony comes in the second section with its abrupt key-shift, and Caesar follows with a smoothly colloquial lyrical transition:

> Nobody near us
> to see us or hear us;
> No friends or relations
> on weekend vacations.
> We won't have it known,
> dear, that we own
> a telephone

The whispered intimacy here encircles the lovers in tightening rhymes that culminate in a high-pitched "dear!" In the next section, however, the monotonous rhythms and two-measure imitations return, and the lyric follows with repetitive plans for "raising" a family, naturally a perfectly symmetrical one ("a boy for you, a girl for me") that neatly doubles the balance.

While most songs of the 1920s had negligible verses, the one for "Tea for Two" is not only good but provides a strong contrast to the tiny repetitions of the chorus. There Youmans used long melodic phrases and Caesar matched them with equally long syntactic structures:

> I'm discontented
> with homes that are rented
> so I have invented
> my own.
> Darling, this place is
> a lover's oasis,
> where life's weary chase is
> unknown!

In further contrast to the flat *me* and *you* masculine rhymes of the chorus (where the rhymes fall on a single accented syllable), Caesar laced the verse with clever feminine rhymes, "pl*ace is*"/

"*oasis*"/"*chase is* (two-syllable rhymes where the second syllable is unaccented).

While he continued to crank out hits for decades, Caesar's long career flourished between 1919, when he put lyrics to George Gershwin's biggest-selling song, "Swanee," and 1928, when he set Joseph Meyer and Roger Wolfe Kahn's "Crazy Rhythm." Musically and lyrically, the latter song derives from George and Ira Gershwin's "Fascinating Rhythm" (1924); while the music is less complex than Gershwin's, relying on a repeated two-measure fragment, its monotony better suits Caesar's lyrical motif of nervous addiction:

> What's the use
> of Prohibition?
> You produce
> the same condition,
> crazy rhythm,
> I've gone crazy too.

Like so many other lyricists of the decade, Caesar heightens the musical repetitions with insistent rhymes (*use*/pro*duce*, *you*/*too*), alliteration (*prohibition, produce*), and even uses the musical accent to create a faint off-rhyme between r*hythm* and *ition*. The character who emerges from the lyric is a '20s variant of the helpless victim of love—a doped-out addict of the feverishly repetitive rhythms of the jazz age.

Perhaps the master at fitting verbal shards to the rickey-tick music that set the nervous pace of the 1920s was Gus Kahn, the only lyricist to be honored with a Hollywood "bio-pic," *I'll See You in My Dreams*. Kahn's knack with those oscillating melodies was demonstrated one day when he was visiting Eddie Cantor's house. Idly picking up one of little Margie's toys, a mechanical pig, Kahn wound it up and watched it waddle across the floor. On the spot, so the story goes, he spontaneously improvised a lyric that matched the pig's herky-jerky movements:

> Yes, Sir! That's my baby!
> No, Sir! Don't mean maybe!

Like Caesar, Kahn wrote not only with alacrity but with versatility; in the course of his career, according to Warren Craig, Kahn "worked with more successful composers than any other lyricist."[4]

His first major success came in 1915, when he teamed with an old-time Alley composer, Egbert Van Alstyne, to produce the cloying "Memories." It was a song very much in the turn-of-the-century mode, its arching melody reminiscent of Van Alstyne's earlier hit, "In the Shade of the Old Apple Tree" (1905), and Kahn matched its musical opulence with long, open vowels, dispensing the *o* and *e* of "memories," generously across the chorus.

In 1921, however, Kahn showed that he was equally adept with the musical style of a new decade. Given the repeated four-note musical phrases of Raymond Egan and Richard Whiting, Kahn hit his lyrical stride:

> Every morning
> every evening
> ain't we got fun?
> Not much money,
> oh, but honey,
> ain't we got fun?

Not only does Kahn use abrupt, colloquial—even ungrammatical—phrases, he abandons syntax for the telegraphic connections of conversation. Truncated slang phrases like "Not much money" are the verbal equivalent of the syncopated musical fragments, and Kahn heightens the ragged feel of both music and words with rhymes that come off and on the beat—*got* against *not*, *fun* against *money* and *honey*. The playfulness of both music and lyric surprisingly drops a proverbial rhyme in a winking celebration of the sexual compensations of poverty:

> There's nothing sure-
> r: the rich get rich
> and the poor
> get children

Kahn uses alliteration to underscore his innuendoes, the *ch* of ri*ch* begetting *ch*ildren, though rhyme can also coyly serve his ends:

> in the meantime
> in between time

Here Kahn lets the very repetitiveness of musical and lyrical phrasing suggest the incessantly frenetic "fun" itself. It's not surprising that Fitzgerald has Jay Gatsby nervously request this song to celebrate his reunion with Daisy.

In 1922 Kahn began a long collaboration with Walter Donaldson, whose music carried the repetitive style of the period to its extreme. Donaldson frequently built his melodies around the insistent device of moving back and forth between two notes, and Kahn had to devise various lyrical strategies to offset the monotony. In "Carolina in the Morning" (1922), for instance, he drawled out off-rhymes on f*iner* and Caro*lina* to get out of the musical rut, but in the release, when Donaldson varied the melody, Kahn maintained the to-and-fro pattern in the lyric with incessant internal rhymes:

> where the *mor*-
> ning gl*or*-
> ies
> twine around
> the d*oor,*
> whispering pretty st*or*-
> ies
> I long to hear once *more*

When Donaldson's music returns to its repetitive pattern, Kahn shrewdly dampens his rhymes with alliterating *d*'s and *l*'s: "If I had Aladdin's lamp for only a day."

Gus Kahn had still more herky-jerky novelty hits with Donaldson, like "Yes Sir, That's My Baby," and, with Ted Fiorito, "Charley, My Boy," but he was also alert to the shift toward "sob ballads" signaled by Irving Berlin's "All Alone" and What'll I

Do?" Collaborating with bandleader Isham Jones, Kahn turned out three romantic laments in 1924, and in each he cleverly let the repetitiveness of the music suggest the motif of fate. The best of these, and, according to Johnny Mercer, the greatest popular song ever written, was "It Had to Be You." Mercer never explained why he singled out this song, but perhaps it was because of the way Kahn refreshed that oldest of romantic figures—one Mercer himself often refurbished—the lover trapped by fate (or, in Mercer's case, "That Old Black Magic").

The catch-phrase sets a nonchalantly resigned tone that perfectly matches Isham Jones's casual melody: the prosaic "It had to be . . ." fits the to-and-fro oscillations between the notes D and E, while the lyrically wry "you" rhymes back on *to* but falls, surprisingly, on an F sharp, an interval Wilder praises as "by no means typical of the twenties."[5] Then, after the "it had to be" formula is repeated, "you" falls on the even more unusual G sharp. The musical pattern is the standard thirty-two measures divided into four eight-bar sections, but rather than breaking his lyric into the usual four separate segments, Kahn works against, rather than within, the musical divisions. The first section, for example, closes with a dangling pronoun:

> I wandered around
> and finally found
> the somebody who

Only in the next section does "who" get a verb and a rhyme—the first instance, I believe, where an Alley lyricist stretched syntax *across* the boundaries of the four-part musical structure.

Such elastic syntax is yet another strategy for subverting musical repetition and gives the lyric a slangy drive that makes even fatalism sound buoyant:

> could make me be true,
> could make me be blue
> and even be glad
> just to be sad
> thinking of you

At the end of the lyric Kahn intensified this colloquial ease by dropping his "of's" and "it's":

> with all your faults,
> I love you still.
> It had to be you,
> wonderful you,
> had to be you.

Kahn here alludes all the way back to Monroe H. Rosenfeld's "With All Her Faults I Love Her Still" (1888). But the allusion, as so often in modern poetry, serves to underscore the gap between old and new, in this case between an old sentimental Alley ballad, with its pompous diction and insufferably patronizing lover, and Kahn's own upbeat 1920s victim of fate, who laments his helplessness with slangy insouciance. The title phrase so perfectly suits this modern lover's posture that Tom Adair borrowed it from Kahn in 1941 when, in "Everything Happens to Me," he had that same world-weary figure complain,

> I fell in love just once,
> and then it had to be with you

Kahn's two other collaborations with Isham Jones in 1924 produced "I'll See You in My Dreams," whose jaunty title is belied by the pathos of the lyric. Much better is the harshly alliterative complaint, "The One I Love Belongs to Somebody Else":

> The hands I hold
> belong to somebody else.
> I'll bet they're not so cold
> to somebody else!

Central to the harsh consonantal play, "belong" also carries over the rhymes on *songs* and *strong* from the preceding section. "Else," on the other hand, sets up slant rhymes in the following section on sh*elf* and yours*elf* and, even more faintly, on *alone* and *fall:*

> It's tough to be alone on the shelf;
> it's worse to fall in love by yourself

Even the normally soft consonants, *f, l,* and *s,* work with the slang
formulas "I'll bet . . . ," "It's tough . . . ," "It's worse . . ." to give
this victim of fate a hard-boiled heartache.

Collaborating again with Walter Donaldson in 1928 on
songs for Ziegfeld's musical, *Whoopee,* Kahn found a new match
between his favorite motif of fate and the repetitive musical style
of the decade. "Makin' Whoopee" was a phrase coined by Walter
Winchell, and Kahn cleverly exploits its sexual connotation by
counterpointing the uniqueness of the phrase itself against the
commonness of the act it suggests. Using Donaldson's to-and-fro
phrasing, Kahn's lyric takes a winkingly cynical view of sex as
equally repetitive:

> Another bride, another June,
> another sunny
> honey-
> moon,
> another season,
> another reason,
> for makin' whoopee.

The singer casts an ironic eye on the marriage he witnesses,
noting the discrepancy between his view of the redundancy of
the ceremony, itself made up of repeated phrases ("I do," "I do"),
and its unquestioned uniqueness to the participants. When the
nervous groom "answers twice," moreover, he adds to the repetitiveness of both lyric and music.

For the first year of marriage that stammering groom takes
up repetitive chores, which Kahn describes in rhymes and a so/
sew pun: he's "washing d*ishes* and baby cl*othes*" and is "*so* ambi-
tious" he even "*sews*." Even so, after just a year,

> she feels neglected
> and he's suspected
> of makin' whoopee

In the wake of such double rhymes, Kahn can even make single
rhymes seem to repeat themselves:

> she sits alone
> most ev'ry night
> he doesn't phone her,
> he doesn't write

By bracketing the masculine rhyme for "alone" in "phone her" Kahn manages to make it sound feminine, much like the husband who used to wash dishes and sew. The pattern culminates in the Berlin-like folded-over rhymes that trace his completion of the cycle from marriage to adultery:

> he says he's "busy"
> but she says "Is he?"
> He's makin' whoopee.

Kahn deftly underscores his theme of unchanging change by shifting the meaning of his title phrase—from marriage festivities to reproduction to adultery—even as he repeats it.

In "Love Me or Leave Me," also from *Whoopee*, Kahn gave yet another witty twist to the figure of the fated lover, who now delivers an ultimatum to a jealous mate. Once again, Kahn skillfully uses Donaldson's insistent, repeated phrases to create a faceted verbal mosaic whose theme itself is monotonous oscillation.

> Love me
> or leave me
> and let me
> be lonely
> you won't
> believe me
> and I love
> you only

Kahn intensifies the repeated musical fragments with overlapping alliteration (*leave-love-be*l*ieve*), assonance (*lonely-won't-only*), and with the grammatical byplay of "let" me and "leave" me that gets a further twist with "be*lieve* me." An even more cleverly faceted sequence,

> You might
> find the night
> time the right
> time for kissing
> but night
> time is my t-
> ime for just
> reminiscing

culminates in a buried rhyme—*might, night, right, my t*(ime).

While repetitive, Donaldson's music nevertheless swerves up and down within the octave and even jumps keys, and Kahn's lyric follows with sharp vacillations between plea and ultimatum. At the highest reach of the release Kahn comes up with a paradoxically helpless boast:

> I intend
> to be
> independ
> ently
> blue

Here again Kahn rings his characteristic changes on the figure of the lover as defiant victim, oscillating, with the music, between soaring assertion and repetitive moaning that stretches "blue" over three notes.

As the 1920s drew to a close, Kahn found a new home for his lyrical style in films. After a final Broadway musical in 1929, *Show Girl* (written with George and Ira Gershwin and featuring such songs as "Liza"), Kahn worked in Hollywood until his death in 1941. There, as Warren Craig notes, he wrote for more than fifty motion pictures and turned out such classics as "Thanks a Million," "San Francisco," "Love Me Forever," and "You Stepped Out of a Dream."[6] In 1933, with Edward Eliscu and Vincent Youmans, Kahn wrote the songs for the first Fred Astaire–Ginger Rogers film, *Flying Down to Rio;* numbers such as "Orchids in the Moonlight," "Music Makes Me," and, especially, "The Carioca" launched the film career of the great dance team and set the standard for the finest Hollywood musicals of the era. The following year, in collaboration with Victor Schertzinger, Kahn wrote songs, including the hit title song, for *One Night of Love*—the first motion picture to receive an Academy Award for its score.[7]

While many of his later lyrics, such as "Dancing in the Moonlight" (1934), were set to repetitive melodies similar to those of the 1920s, Kahn could also adapt to the newer musical idioms of the 1930s. For the less repetitive, more driving, jazz, blues, and swing styles, he used the same rough slang that newer lyricists such as Dorothy Fields and Ted Koehler adopted in their Cotton Club songs. In "Dream a Little Dream of Me" (1931), for example, he toyed with clipped, telegraphic phrasing, and in "I'm Thru With Love" (1931) he took an angry catch-phrase title—more biting than "It Had to Be You" or even "Love Me or Leave Me"—and carved out a new version of his perennially victimized lover, one who wittily boasts, "I have stocked my heart with icy frigidaire."

Along with Irving Caesar and Gus Kahn, other lyricists of the 1920s found striking lyrical matches for the repeated-note, repeated-phrase pattern of the decade's music. In 1924 Walter Hirsch created a minimalist lyric to fit Fred Rose's spare and simple musical oscillations:

> Do I
> want you?
> Oh my,
> do I?
> Honey,
> 'deed I do.

By building repetitive, short phrases around the same long vowels, Hirsch creates a baby-talk marriage vow with an understatedly passionate edge, readymade for the coy little-girl singers of the era such as Ruth Etting.

Like the songs, such flappers were the embodiment of erotic

petiteness, and in 1925 Sam Lewis and Joe Young celebrated
those aesthetic features even as they employed them:

> Five foot two
> eyes of blue
> but oh
> what tho-
> se five foot could do!

Their sharp turns of phrase, bad grammar, and genteel French—
all stitched together by embedded *o* rhymes—mirror the faceted
catalog of flapper beauties:

> turned-up *no*se
> turned-down *ho*se
> never had *no*
> other b*eau*s

From such neatly doubled negatives, these tiny repetitions finally
dissolve into baby-talk babble:

> could she
> could she
> could she coo?

Just where the fragments are most minimal, they blur into
"kootchy-coo," the erotic lingo of the "boop-boop-a-doop" girls.
   Sometimes lyricists undercut both musical and lyrical repetitions by taking the standard Alley gimmick of repeating the title
phrase at the beginning and end of the chorus but using it to give
the phrase a different meaning. Thus in 1929 Roy Turk opened
a chorus with

> Mean to me,
> why must you be
> mean to me?

but closed with

> Can't you see
> what you mean to me?

Even though the catch-phrase title, in standard Alley-fashion, appears at the end of the song, its meaning has been completely reversed.

Less radical a turnaround of a title phrase was Edward Eliscu and Billy Rose's variations on "More Than You Know" (1929). Initially, it prefaces a declaration of undeclared passion:

> more than you know,
> more than you know,
> man of my heart I love you so

But then it serves as an understated afterthought:

> lately I find
> you're on my mind
> more than you know

Finally the phrase, slightly modified, reprises its original function of unexpressed emotion: "more than I'd show, more than you'd ever know."

Repetition lent itself as easily to high spirits as it did to heartache. In "Bye, Bye Blackbird" (1926), for example, Mort Dixon used the insistent quarter-note rests in Ray Henderson's melody to clip a syllable from his lyrical lines and create tersely insouciant resolve:

> Pack up all my
> care and woe ( )
> here I go ( )
> singing low ( )

When the melody shifts to languorous whole and half-notes, Dixon slides into the long vowels of his title-phrase but keeps his repetitive pattern going by rhyming *pack* with *black*bird, *my* with *bye*. In the release, the music turns to longer phrases, and Dixon opens up his tight-lipped lyric yet uses internal rhymes to continue his verbal faceting:

> No one here *can* love *and* understand me,
> Oh what hard luck stories they all hand me

In the final A section he returns to telegraphic understatement,

> Make my bed and
> light the light (   )
> I'll arrive (   )
> late tonight

but recaptures some of the emotion of the release with "late tonight" since the music of that measure varies the repetitive pattern by dropping the expected quarter-rest. By placing such an elongated (but still three-syllable) phrase at this musical point, Dixon gives "late tonight" a slyly intimate overtone.

Repetition, as one might expect, was especially amenable to the symmetry of duets, and in the best of these, Harry Woods' "Side by Side" (1927), the tight parallels in verbal and musical phrasing mirror the closeness celebrated in the lyric itself. Woods uses slang formulas ("Oh we ain't got . . . Maybe we're . . .") and refreshes clichés by musical ragging to create affectionate understatement:

> When they've all had their quarrels and par-
> ted,
> we'll be the same as we star-
> ted,

Woods underscores the contrast between "we" and "they" syntactically by shifting from past tense verbs to ongoing participles: we're "travelin' along, singin' a song."

Such symmetry lent itself equally well to themes of solitude, much as it did in Irving Berlin's faceted ballads of the '20s. The sparse rooms depicted in "All Alone" and "All by Myself" mirror the confines of musical phrasing, yet these same solitary interiors could turn deliciously intimate in other lyrics that followed box-like melodies:

> a turn to the right,
> a little white light

Here in "My Blue Heaven" (1927) George Whiting intensifies the intimate enclosure of Walter Donaldson's repetitive melody with cloying *i* rhymes and overlapping alliteration.

Probably the wittiest transformation of this box-like room came in Andy Razaf's lyric for Fats Waller's "Ain't Misbehavin' " (1929). Razaf reverses the formula by having his solitary singer rejoice to be "home about eight" and alone in his room—"just me and my radio"—since such isolation is a sign that he is happily faithful to his absent lover. As the singer celebrates that lonely fidelity with increasing fervor, the space of the tiny room constricts even further:

> like Jack Horner,
> in the corner,
> don't go nowhere,
> what do I care?

Here the repeated telegraphic slang phrases breathe an exuberant air into the lonely room of a Berlin sob-ballad.

Of all the masters of musical and lyrical repetition during the 1920s, no songwriting team produced more hits than the threesome of Buddy DeSylva, Lew Brown, and Ray Henderson, yet Alec Wilder includes not one of their songs in his extensive survey of popular music. While the omission may be due to the monotony of Henderson's music, DeSylva and Brown should certainly be celebrated for the rich variety of verbal mosaics they fitted to that confining style. Both lyricists had a long apprenticeship on the Alley before the partnership formed. Lew Brown had been writing since 1912, frequently collaborating with old-time composers like Albert Von Tilzer on everything from "state" songs to "girl's name" songs—and even mixing the two genres in "Kentucky Sue" (1912).

Buddy DeSylva had worked the different but equally well-worn vein of ten-cent philosophy songs, matching Jerome Kern's opulent melodic line in "Look for the Silver Lining" (1920) with such bloated platitudes as "A heart full of joy and gladness will always banish sadness and strife," then soaring even higher in "April Showers" (1921):

> And where you see clouds
> upon the hills
> you soon will see crowds
> of daffodils

Al Jolson was so fond of this watered-down Wordsworth that he would solemnly recite the verse as if it were the profoundest poetry. DeSylva could use musical repetition in distinctly different ways, however, such as underscoring lyrical double entendre:

> If you knew Susie
> like I know Susie
> Oh! Oh! Oh what a girl!

Here the repeated *o* and *e* rhymes, counterpointed in *knew* and *know,* fill the suggestive musical gap between Su*sie* and *oh! oh! oh!* with all the naughty nuances of carnal knowledge, nuances drilled home by Eddie Cantor's rolling eyes.

In the early 1920s DeSylva was paired with George Gershwin on songs for George White's annual *Scandals.* In 1922 he collaborated with a novice, Ira Gershwin, on George's rangy, blue-note studded "I'll Build a Stairway to Paradise," and, in 1924, DeSylva and Ballard MacDonald had another hit with Gershwin's more typically '20s repetitions in "Somebody Loves Me." It was when George Gershwin left the *Scandals* in 1924 (to team with brother Ira on *Lady, Be Good!,* the first of their musical comedies) that George White created an unusual three-man songwriting team by uniting DeSylva with Lew Brown and Ray Henderson.

> The first two were lyricists, the third, a composer. Their first score was for the 1925 edition of [George White's *Scandals*]. For the next half dozen years or so, the three men worked so intimately and harmoniously that it was not always clear where the work of one ended and that of the other two began. There were times when the composer, Ray Henderson, helped to write lyrics, and when the two lyricists provided ideas to the

composer. . . . Consequently, in talking about the songs of
DeSylva, Brown, and Henderson it is necessary to speak of
them as a single creative entity.[8]

One hallmark of that threesome, seen in one of their first hits in
1926, was their use of extended poetic images, sometimes strik-
ingly surreal ones, to offset the brief repetitive musical phrases:

> From a whippoorwill,
> out on a hill,
> they took a new note,
> pushed it through a horn,
> till it was worn,
> into a blue note

The extended obstetrical metaphor of "The Birth of the Blues,"
from the *Scandals* of 1926, implicitly equates trumpet and fallo-
pian tubes and even gives "blue" a babyish twist.

Just as George Gershwin aspired to greater musical sophistication when he left George White's revues for the higher plateau
of musical comedy, DeSylva, Brown, and Henderson attempted
to write more sophisticated songs when they worked on satirical
musicals like *Hold Everything!* (1928). In the hit of that show,
"You're the Cream of My Coffee," for example, they tried to set
Henderson's musical repetitions to the rigors of a "list" song—a
lyric that consists of a catalog of witty images. List songs only
succeed, however, when each successive image "tops" the preceding one, and thus such songs are usually attempted by only the
most skillful theater lyricists (the archetypal list song is, of course,
Porter's "You're the Top"). Measured against that standard,
"You're the Cream in My Coffee" falls short, precisely because
the best images come first:

> You're the cream in my coffee,
> you're the salt in my stew

Such a witty metaphor, we recall, struck even Cleanth Brooks as
the same sort of startling "conceit" that abounds in modern
poetry.

DeSylva and Brown, moreover, use such extravagant metaphors to create the qualities Brooks praises in modern poetry:
paradox, "ironical tenderness," and "a sense of novelty and freshness with old and familiar objects."[9] The utterly ordinary comple-
ments of salt and cream serve as extraordinary love compliments,
and they in turn refresh the tiredest of romantic staples, the
catalog of the beloved's beauties—a device already stale when
Shakespeare wrote "My mistress' eyes are nothing like the sun."
Not only are DeSylva and Brown's metaphors wittily understated, they turn paradoxical as well in the lines that follow:

> you will always be
> my necessity,
> I'd be lost without you

Cream in coffee, salt in stew, are, technically, not necessities at all
but supplements, yet, somehow, paradoxically necessary addi-
tives for those who use them. The imagery is matched by equally
clever sound play. The tritest of rhymes, the *ee-oo* pair that under-
lies everything from "Tea for Two" to "Sweet Sue," is enlivened
by mating *me* and *you* to cof*fee* and st*ew*. Interwoven among the
lines are alliterative threads that tie *salt, stew,* and *necessity* together; then *lost* not only completes the pattern but nearly
metathesizes *salt.*

Having established two witty conceits, however, DeSylva and
Brown still faced twenty-four bars of music. While an Imagist
poet could turn out a pair of images,

> The apparition of these faces in the crowd;
> Petals on a wet, black bough.

then stop, a lyricist had to come up with not only more images
but increasingly better ones. In the second A section of "You're
the Cream in My Coffee," the strain begins to show:

> You're the starch in my collar,
> you're the lace in my shoe.

Although they shift from one "necessity"—food—to the mainstay of clothing, the metaphors are less commonplace and thus less striking. DeSylva and Brown do maintain the alliterative pattern, and may even be suggesting a pun on *lace* (as what cream, salt, and starch all do to what they're added to), but the strain is even more apparent in the release:

> so, this is clear, dear,
> you're my Worcestershire, dear

While the sound play in "Worcestershire" is wonderful, the strategy of transforming the prosaic into the extraordinary is lost. By the final A section, Brown and DeSylva don't even seem to be trying:

> you're the sail of my love boat,
> you're the captain and crew

Abandoning even the "necessity" structure (which would lead them to shelter after food and clothing), they settle for the very sort of flat cliché the song started out by mocking.

They did better when they laced their lists with the suggestive imagery that was their trademark. "Button Up Your Overcoat" (1928), for example, is a duet, where the girl's chorus proffers a list of images that initially smack of maternal care—"Eat an apple every day," "Be careful crossing streets"—but then switch to deadpan flapper advice: "Get to bed by three" and "Stay away from bootleg hooch." In the boy's chorus the images are even more slyly suggestive:

> Wear your flannel underwear
> when you climb a tree

The same concern for the beloved's treasured parts spawns a catalog of images that symbolically spell out the danger to girls who "go out with college boys":

> don't sit on hornet's tails,
> or on nails,
> or third rails

Nevertheless, DeSylva and Brown don't trust the suggestiveness of their images alone but punctuate each with an "oo-oo! oo-oo!" that matches Henderson's insistently repetitive music.

In "(Keep Your) Sunny Side Up" (1929), one of their last hits as a threesome, they elaborated a list of their most surreally suggestive imagery:

> Keep your sunny side up, up!
> Hide the side that gets blue.

The vaguely risqué advice, a refreshing transformation of the pompous didacticism of "Look for the Silver Lining," invokes a two-sided figure one might see in a Dali painting:

> Stand upon your legs,
> be like two fried eggs!

"Sunny Side Up" was written not for a *Scandals* revue or a Broadway musical but as the title song for a movie. Ironically, it was the emergence of talking pictures with Jolson's *The Jazz Singer* in 1927 that at first galvanized, then dissolved, the song-writing team. Al Jolson commissioned them to write songs for his 1928 movie, *The Singing Fool*, but DeSylva, Brown, and Henderson should have sniffed trouble when a song they wrote as a parody of Tin Pan Alley sentimental fare, "Sonny Boy," was sung "straight" by Jolson and became a stupendous hit.

Writing such songs for films was far more profitable than writing for the stage, and by the end of the 1920s, according to Sigmund Spaeth, "the screen actually offered more popular music than the stage."[10] Tin Pan Alley songwriters began heading west, and the trio of DeSylva, Brown, and Henderson were among the first to strike gold in the Hollywood hills. One of their hits made the new medium itself the subject of a love song. "If I Had a Talking Picture of You" presents the typical '20s singer alone in a barren room, but in place of the standard Berlin furnishings of telephone and photograph, this woebegone lover envisions his own home movies:

I would sit there in the gloom
of my lonely little room
and applaud each time you whispered,
"I love you! I love you!"

The conceit also allowed them to give a new turn to the old 1920s lover—paradoxically helpless yet boastful:

I would give ten shows a day
and a midnight matinee

The joys of voyeurism here make the artificial girl more appealing—and acquiescent—than the real thing.

Hollywood was so hungry for the work of DeSylva, Brown, and Henderson that Warner Brothers bought out their entire catalog of songs early in the Depression. The extravagant sale, it seems, contributed to the dissolution of the team, for DeSylva soon gave up songwriting to become a producer of such box office successes as the Shirley Temple films. He also produced Broadway shows, such as *Anything Goes,* and even showed in *Take u Chance* (1932) that he could still turn out a witty theater lyric, such as Ethel Merman's naughty narrative, "Eadie Was a Lady," and the sensuously imagistic "You're an Old Smoothie":

You're an old smoothie,
I'm an old softie,
I'm just like putty
in the hands of a girl like you.

Such metaphoric play was all but banned in Hollywood, which in the early days of film musicals wanted its songs to be blander than the blandest Alley fare. Thus, while Brown and Henderson continued to collaborate for a few more years, they gravitated back to Broadway, where theater audiences were more appreciative than the movie moguls of such metaphysical conceits as "Life Is Just a Bowl of Cherries" (1931). Such audiences, however, also had come to expect the sophistication of Hart, Gershwin, and Porter, and Brown and Henderson were no match for that standard or for newer teams like Howard Dietz and Arthur Schwartz or Yip Harburg and Harold Arlen. After a couple of unsuccessful musicals in the early '30s, Brown and Henderson went their separate ways.

The dissolution of DeSylva, Brown, and Henderson resembled the fate of other Alley songwriters of the 1920s. While their bone-simple lyrics, set to repetitive music, found a new home in films, the songwriters themselves often did not. What the studios wanted, as Ethan Mordden argues, was practically a parody of Tin Pan Alley songs of the '20s—songs so simple, formulaic, and "universal" that they could be turned out "cookie-cutter fashion." Lyricists like DeSylva and Brown, who had learned to stamp the repetitive melodies with their own "peculiar, individual edges," were too sophisticated for films that called for songs free of the "penetration of individuality."[11] Given censorship pressures on the movie industry in the early 1930s, moreover, lyrics that smacked of double entendre were taboo.

Yet Hollywood was only one factor in the demise of the Alley's heyday. Even before the decade had begun, younger lyricists like Lorenz Hart were chafing against the banality of popular songs, and when his intricate and sophisticated lyric for Richard Rodgers' "Manhattan" became a surprising popular success in 1925, a new era of wit began, an era when, for the first time in American popular song, people began *listening* to the lyric. Hart, along with Ira Gershwin and Cole Porter, wrote almost exclusively for the Broadway theater, yet their love songs were aimed at Tin Pan Alley's popular market. The success of their lyrics, as we will see, set a standard that few Alley wordsmiths could match.

Funny Valentine: Lorenz Hart

If it hadn't been for Larry Hart, none of us would have felt free to write colloquial lyrics. He took the way people talked and put them into lyrics. That doesn't mean much, but in the early twenties nobody had ever done it,

OSCAR HAMMERSTEIN

In 1919, when Richard Rodgers first met him, Lorenz Hart was wearing tuxedo trousers and a bathrobe—a clash of formal and casual styles that would fashion his best lyrics. It was of rhyme, Rodgers recalled, that Hart spoke, castigating Tin Pan Alley lyricists for their failure to use "interior rhymes, feminine rhymes, triple rhymes and false rhymes"—indeed anything but the simplest and tritest "juxtapositions of words like 'slush' and 'mush.' "[1] Only P. G. Wodehouse was exempt from Hart's wrath, and when he found that Rodgers shared his enthusiasm for the Princess Theatre shows, the twenty-three-year-old lyricist and sixteen-year-old composer spent the afternoon listening to the songs of Wodehouse and Kern on Hart's victrola. "I left Hart's house," Rodgers said, "having acquired in one afternoon a career, a partner, a best friend, and a source of permanent irritation."

Hart's addiction to rhyme, glimpsed in this first meeting with Rodgers, stayed with him until his death, from alcoholism, in 1943. When he subordinated that resounding urge to melodic contours, he disclosed brilliant rhymes within the flow of conversational speech. Oscar Hammerstein, who replaced Hart as Rodgers' collaborator, praised Hart not only for his "rhyming grace and facility," admitting he "would not stand a chance" against Hart "in the field of brilliant light verse," but also for his ability to contain such rhymes within perfectly colloquial diction."[2] When Hart's addiction to rhyme overran vernacular bounds, however, he was criticized by fellow-lyricists such as Howard Dietz, who quipped: "Larry Hart can rhyme anything— and does!"

Hart defended himself against the charge that "all I could do was triple-rhyme" by citing a lyric where he passed up the chance to rhyme in order to stay colloquial: "Now just take a look at this lyric: 'I took one look at you, that's all I meant to do, and then my heart stood still.' I could have said, 'I took one look at you, I threw a book at you,' but I didn't." In the example Hart cites, it was not only his own penchant for rhyme that had to be resisted but the temptation proffered by Rodgers' music, which repeats the same melodic phrase for "I took one look at you" a few intervals apart for "that's all I meant to do" with the high note of each parallel phrase on "look" and "meant"—an open invitation for a thumping rhyme.

Howard Dietz's own collaborator, composer Arthur Schwartz, also defended Hart, noting that Rodgers' melodies frequently dictated relentless rhyming by their "patterns and schemes of construction."[3] One such pattern that Rodgers "continued to use throughout his career" was "that of returning to a series of notes, usually two, while building a design with other notes."[4] In the early "Blue Room" (1926), for example, Hart matched this two-note device with a sequence of feminine rhymes that ranged from the prosaic—"blue room/new room/ for two room"—to the pyrotechnic:

> You sew
> your trousseau,
> and Robinson Crusoe . . .

Even in their last hit, "Bewitched" (1940), Hart continued to refresh Rodgers' musical oscillations, this time with flippant, but still thoroughly conversational, rhymes:

> vexed again,
> perplexed again,
> thank God I can be oversexed again!

Not only did Rodgers' music provide a structure for such intricate rhymes, its sonority (Cole Porter once quipped that every Rodgers melody had a certain "holiness" about it) served as a perfect counterpoint to Hart's cynical urbanity. Rodgers himself said that the secret of their best songs was the clash between a "sentimental melody and unsentimental lyrics," a clash punctuated by caustic rhymes.[5]

Because they complemented each other so perfectly, Rodgers and Hart wrote songs that, in their blend of word and music, rival those of Irving Berlin or Cole Porter, who wrote both words and music, as well as songs by those perfectly matched brothers, Ira and George Gershwin. It took a number of years after their first meeting, however, for Rodgers and Hart to perfect that blend, and longer still for it to be accepted, not only by Broadway theater audiences but by the mass popular market of Tin Pan Alley. Between 1919 and 1925, the fledgling team made so little headway that Rodgers nearly abandoned songwriting to go into the children's underwear business.

At first, though, it had seemed success would be instantaneous. Teamed with a young librettist named Herbert Fields, Rodgers and Hart sold one of their first songs, "Any Old Place With You," to Herbert's father, the famous vaudevillian Lew Fields, who interpolated it into his 1919 production *A Lonely Romeo.* A forerunner of "Manhattan," "Any Old Place With You" was a geographical list song that did well in the show. Some of the place-names rhyme with colloquial ease ("I'd go to hell fer ya—or Philadelphia" got a big laugh), but others invert the conversational flow to force the rhymes:

> In dreamy Portugal,
> I'm goin' to court you, gal.

By the time we reach "I'm goin' to corner ya in California" the rhymes have become monotonous and predictable.

"Any Old Place With You" got Rodgers and Hart nowhere. Although he had promised to let them write all the songs for his next show, Lew Fields turned instead to the established team of Alex Gerber and Sigmund Romberg, and, for the next few years, Rodgers and Hart had to settle for writing songs for amateur musicals, such as the Columbia Varsity shows. They couldn't even get to Tin Pan Alley when Tin Pan Alley came to them. Arthur Schwartz, who worked at Brant Lake summer camp where Hart served as dramatic counselor, recalled how one day in 1920 the famous music publisher Elliot Shapiro came to visit and agreed to hear some Rodgers and Hart songs:

> Now, Elliot Shapiro was not the man with the cigar that you expected to come out of Tin Pan Alley. (I'd seen them look like that in the movies.) He was dignified and grim-faced, and had never heard of Larry Hart or the very new and very young team of Rodgers and Hart. We went to the recreation room (it also doubled as gymnasium), where there was a piano. Mr. Shapiro's face went from grim to grimmer as Larry sang and I played four or five songs. He shook his head. "It's too collegiate. You fellows—I think he thought that I was Rodgers—"will never get anywhere unless you change your style. These songs are great for amateur shows. I've got to tell you the truth, fellas. Change your style or give up."[6]

Shapiro was right. Hart's lyrics for the Columbia Varsity Show of 1920 could interest only the alumni, and his place-name rhymes again relied upon poetic inversions and verbal filler:

> Bulldogs run around New Haven,
> Harvard paints old Cambridge red,
> and even poor old Philadelphia
> really has a college, it is said

Even a reviewer for the *Columbia Spectator,* while praising the show for its "catchy music," found the best lyric to be not one of Hart's but an interpolated song by another Columbia student— Oscar Hammerstein.

What Elliot Shapiro said of Rodgers and Hart songs in 1920—"too collegiate"—he might also have said of Broadway itself a few years later. "The early twenties," recalled actress Edith Meiser, "brought the first wave of people from colleges entering the theater."[7] She herself was part of a troupe of recent graduates who wanted to do a small revue at the Garrick Theatre to raise money for the Theater Guild. The *Garrick Gaieties* was one of the first "sophisticated" revues of the mid-1920s, as much of a streamlined antidote to Ziegfeld extravaganzas as the Princess shows had been to florid European operettas. Even more loosely constructed than the Princess musicals, the new revues aimed at the self-styled "smart set" of the 1920s, an audience that wanted urbanely witty fare.

Aiming to please, the "college kids" strung together satirical sketches on current Broadway shows, as well as the Guild's own avant-garde forays into O'Neill and Ibsen. What they needed were songs, and when they heard about a young composer named Richard Rodgers, Meiser called on him. He played her numbers from Columbia Varsity Shows and other amateur productions; she was unimpressed. Then he played a new song, "Manhattan," from an unproduced show of 1922 called *Winkle Town,* and, as she put it, "I flipped."

"Manhattan" made an overnight success of Rodgers and Hart and turned the 1925 *Garrick Gaieties,* originally planned for only two benefit performances, into a long-running Broadway hit. Robert Benchley praised it as "the most civilized show in town," and another denizen of the Algonquin Round Table, Alexander Woollcott, dubbed it "bright with the brightness of something new minted."[8] While Woollcott praised Hart's witty lyrics, however, he expressed doubts about how "singable" they were, doubts that there could ever be a true marriage between society verse and popular song.

"Manhattan" quickly dispelled those doubts. Almost immediately it became an independent hit through Tin Pan Alley's sheet music and record sales. Moreover, it was the first popular song whose lyric literally made headlines. Newspapers reprinted the "contagious lyrics," the *Evening Graphic* (May 20, 1925) under the headline "Rhymes That Do!" The *Morning Telegraph* (October 4, 1925) praised "Manhattan" as a "New York song" whose "unusual lyric" was being reprinted in newspapers across the country and bringing "New York back into the limelight." Even the *Atlanta Constitution* (July 26, 1925) was informing its readers that "Manhattan" was "the most popular piece in New York, you hear it played wherever you go," and advised: "By all means get your orchestras to play "Manhattan" for you, if you wish to dance by 'what is being danced by.' "

In an interview for the *Herald Tribune* (May 31, 1925) Hart confessed his amazement that a lyric with such "very intricate and elaborate rhymes" should be "the hit of the show." "The song hit of the show," he said, "is usually a very simple one with monosyllabic words." "Manhattan" had not only violated that axiom but another as well: "the cleverest lyricists are seldom big-hit writers." Hart cites P. G. Wodehouse as an example of a light verse writer, writing for musical comedy, with "fewer hits" than "many inferior lyric writers." Similarly, he tries to maintain the distinction between Broadway—where the "audience listens" to sophisticated lyrics that never become popular—and Tin Pan Alley—where the "royalty statement of the music publishers" depends upon "banal lyrics." It was a distinction that "Manhattan," with its enormous sheet music sales, had already erased.

"Manhattan" is a list song based on place names, much like "Any Old Place With You," but it is so much wittier than the earlier song that it's hard to believe Hart's claim that its popularity was "not due to any definite improvement in my technique." He takes Rodgers' jaunty, strolling melody and sets it to an urban pastoral. Yet Hart's idyll is an ironic one: "balmy breezes" emanate from the subway and even the ordinarily frenetic pushcarts of Mott Street go "gently gliding by." Using the quick leaps and descents of the melody, Hart can rhyme in blatant Brooklynese:

> The city's clamor can never spoil
> the dreams of a boy and goil,

On the other hand his rhymes can be barely audible, as in Mott *Street's sweet* pushcarts, or the eye and slant rhymes that slide off the title: "We'll have *Man*hatt*an*, the Bronx *and* Stat*en* Isl*and* too."

The real innovation, however, comes from Hart's intricate "ragging" of word against music, using the intervals and pauses of Rodgers' melody to fragment the verbal line. Although similar to verbal "misfits" to music in songs by Irving Berlin and Gus Kahn, Hart goes much further in using Rodgers' musical shifts to split, rather than fit, his lyric:

> what street
> compares with Mott Street
> in July?
> Sweet pushcarts gently gli-
> ding by

Not only does such syllabic fragmentation unearth the *gli/by* rhyme, it rhymes back on Ju*ly* as well, which, in turn, eye-rhymes with "gent*ly*." "The prominence of jazz," Hart told the *Herald* interviewer, "makes things very difficult for the lyric writer" since the music "throws" the rhythm of the lyric "out of gear." But in "Manhattan" he turned that liability into an asset. Extending the principle of fragmentation from rhyme to phrase, he uses Rodgers' abrupt turns, sharp rests, and wide intervals to break up verbal phrases. Thus he first seems to mock the unwashed Village Bohemians in one breath,

> We'll go to Greenwich,
> where modern men itch

but recants in the next:

> to be free;

Such deadpan pausing goes hand-in-hand with a double rhyme as Hart first seems to praise Broadway's schmaltziest hit,

> Our future babies
> we'll take to *Abie*'s
> *Irish Rose.*
> I hope they'll live to see

only to bury it with a caustic rhyme as the music takes a sudden downward turn:

> it close

    The most complex instance of ragging verbal against musical phrasing is a sequence of ambiguously overlapping phrases that rivals the fragmented lines of William Carlos Williams' poems of the 1920s. Williams, for example, uses a simple visual grid of three words/one word to create a complex verbal pattern where "so much depends," as seemingly independent phrases like

> a red wheel

and

> glazed with rain

turn out to depend upon such seemingly independent "lines" as

> barrow

and

> water

or unstable three-word units like

> beside the white

are presented as independent lines, only to be anchored by

> chickens

Using Rodgers' angular musical pauses and repetitions as a similar grid, Hart presents what at first seems a complete statement:

> We'll bathe at Brighton;
> the fish you'll frighten
> when you're in

But the "when you're in" that seems to complete one phrase suddenly also seems to depend upon another:

> your bathing suit so thin

But then that apparent completion seems only to introduce another dependency:

> will make the shellfish grin

And just when we think we're at the end of these overlapping, fish-scale phrases, comes

> fin to fin

"Manhattan" was not the only song from *Garrick's Gaieties* to become popular on Tin Pan Alley. The wittily unsentimental "Sentimental Me (And Poor Romantic You)" was hailed by a reviewer in the *Morning Telegraph* (July 26, 1925) for departing "successfully from the stereotyped 'Moon-June' doggerel of the current love ditties." Like "Manhattan," "Sentimental Me" not only was lauded "by the high-salaried critics" but was a "big money-getter" as well, one that had the "recording managers of the big phonograph companies" all "breaking their necks to put the record on the market."

By the end of the year, *Variety* (December 30, 1925) took official notice of a lyrical renaissance, focussing on Hart (even comparing him to Gilbert), but also praising Ira Gershwin's "great lyrics" for the 1924 production *Lady, Be Good!* (the first successful collaboration of the Gershwin brothers) and Howard Dietz as "another of the newer lyric writers who turns out good stuff." By 1928, Cole Porter, who had been trying for over a dozen years to gain popular acceptance for his witty "list" songs finally had an enormous success with the risqué bestiary "Let's Do It." The golden age was clearly dawning.

For nearly two decades, Hart's lyrics set a new standard for intricately witty rhyming, though from the 1920s to the 1930s one can detect in the lyrics a distinct change in his use of rhyme to bring out the repetitive patterns of Rodgers' music. From rhymes that underscore the euphoria of love, he shifts to ones that create what Anthony Burgess has described as "a certain ambiguity of feeling, wit in the service of frustration or neurosis," yet still without a loss of vernacular ease: "the rhymes do not really call attention to themselves, even at their most ingenious."[9]

Typical of Hart's songs of the 1920s was "Mountain Greenery," the big popular hit from the 1926 edition of *Garrick Gaieties*. A mirror image of "Manhattan," "Mountain Greenery" views the pastoral idyll through urbane eyes. The cosmopolitan lovers again joyously play at Edenic innocence, like the "Mister Omar" they invoke, and regard "nature" as a superb artifice where "God paints the scenery." Triple rhymes like "scenery/greenery" underscore the confected vision (a greenery, after all, is where nature is nurtured) and even mime the transformation of natural product into industrial artifact:

> Beans
> could get
> no keen-
> er re-
> cept-
> ion in a beanery

Here the processing of simple words produces pyrotechnically fragmented rhymes: "keen" rhymes with all but the "s" of "beans," "get" slant rhymes with the middle of "reception," and a bizarre new linguistic hybrid, "-er re-," compounded of a prefix preceded by a suffix, mates with "-ery."

As in "Manhattan," Hart rags words against music, so that a rhyme blurs what should be a syntactic break: "While you love your *lover, let* blue skies be your *coverlet.*" Flashy as such triple rhymes are, the buried ones are even better, like the one on "cook" that is only unearthed when a musical break severs "ing" from "looking." Hart used the two accented half-notes that open the chorus to emphasize two otherwise insignificant words,

> In a

much as Marianne Moore was using a visual syllabic grid to high-light and rhyme articles,

> the
> turquoise sea

and

> an
> injured fan,

as well as even smaller particles of language:

> ac-
> cident lack

or

> dead.
> Repeated

Hart's rhymes are just as hidden until the musical grid accents them: "*In* a moun*tain* greenery."

In addition to the Garrick revues, Rodgers and Hart, along with librettist Herb Fields, wrote a series of Broadway musicals in the 1920s that broke from "the sentimentality that had shackled musi-cal comedy—the soft, even maudlin themes of Victor Herbert's operettas and, for that matter, the hitherto-hallowed effusions of Bolton, Wodehouse and Kern."[10] One sign that Rodgers and Hart had displaced the very team they set out to emulate came from a reviewer who praised them as "the American counterpart of the once-great triumvirate of Bolton, Wodehouse, and Kern."[11]

To sidestep sentimentality in these shows, Hart perfected his clashing rhymes, setting up in "Where's That Rainbow" (1926), for example, a potentially maudlin plea with one rhyme,

> Where's that Lothario?
> Where does he roam,

then debunking it with another, in this case a slickly enjambed wink at Hollywood's Latin lovers:

> with his dome
> Vaselined as can be?

For "Here in My Arms" (1925), he triple rhymed a clichéd "it's adorable" with an unromantically irritated "it's deplorable that you were never there."

The best of these early musicals, *A Connecticut Yankee* (1927), gave Hart a perfect chance to undercut sentimentality by rhyming antiquated diction with modern slang. Like Mark Twain, who mingled western tall-talk with medieval archaisms in the original book, Hart bounced between linguistic extremes in songs such as "Thou Swell." In the verse he gets deadpan clashes, like Twain's, when a compliment that begins "you are so graceful" turns into a burlesque

> you have a face full
> of nice things

Similarly, the highflown "Are you too wistful to care" is deflated by "You're such a fistfull." Even when there is no perceptible clash of diction, a rhyme sequence of poetic inversions,

> Babe, we are well met,
> as in a spell met,
> I lift my helmet

makes "hel-met" sound like an infernal rendezvous (how much defter is Hart's pun than DeSylva and Brown's "plenty of hell-th"). Alec Wilder has praised Rodgers' music for the way "it starts out innocuously" but then "resorts to a series of leaps which fall in odd places."[12] That is also a perfect description of how Mark

Twain told a joke, and of how Hart's wryly skewed rhymes and dictional drops follow the musical contour.

The chorus of "Thou Swell" bubbles over in an ebullient mixture of archaism and slang:

> Both thine eyes are cute, too—
> what they do to
> me.
> Hear me holler I choose a
> sweet lollapaloosa
> in thee!

"Lollapaloosa" is not just a show-off rhyme: it comes from the American tall-tale talk that was Twain's idiom, and, with its connotations of hugeness, works with the expanding imagery—"fistfull" and "face full"—to bring out the literal pun on "swell."

Such "swelling," in turn, works within images of tiny containment:

> I'd feel so rich in
> a hut for two.
> Two rooms and kitchen
> I'm sure would do

Again, "hut" is both American slang yet, given the medieval setting of the song, literally accurate, and the "do/two" rhyme recalls and completes the earlier "cute too/do to" sequence with the third "to/to/two" homonym. The two "twos," back to back, only increase the lyrical "swelling," and as the imagery plays between expansion and contraction, the music alternately climbs up the scale in larger quarter-note and half-note steps, then descends in tiny eighth-note oscillations. In the final B section, Rodgers stops short after four bars of descending pirouettes—but then leaps up to close with four bars of the A melody. Hart marks the abrupt shift by placing the innocuous word "and" at the crucial juncture:

> Give me just a plot of
> not a lot of

land,
AND,
Thou swell! Thou witty! Thou grand!

Such explosive placement not only works deftly with the music but pinpoints the lyrical tension between contraction and expansion—yet does all this with the most casually conversational phrasing: the self-interrupting aside, "not a lot of," suddenly swelling with the addition of the American huckster's expansive "AND. . . ." Anthony Burgess marvels at Hart's ability here "to find rhymes lying around even in colourless morphemes." Even more impressive is that Hart weaves these brilliant rhymes into a thoroughly "conversational declaration which, by the happiest of accidents, has rhymes in it." More impressive still is that Hart takes this blend of artificial rhyme and natural speech and weds it to music. Burgess, in fact, finds that, when the lyric is written out, the rhymes seem "whimsical and overcontrived"; when sung to music, they sound "very natural."

Yet Hart has equally clever—and natural—rhymes that are barely audible, noticeable only when one looks at the words on the page. Lehman Engel has praised Hart's shrewd rhyming of syllables in "Blue Room" (1926), where "future" not only rhymes audibly with "suit your" but subtly sets up the key word "blue" with its first syllable. One could go even further than Engel does and note how that *oo* spills over into "*room*" itself, just as the first syllable of "*hol*iday" quietly sets up a whole sequence of resounding rhymes: "*ball* room," "*small* room," and, curling all the way back in a homonym, "*hall* room."

Some of Hart's rhymes are so unobtrusive that they elude even Engel, who quotes this passage from another *Connecticut Yankee* song, "My Heart Stood Still":

A house in Iceland
was my heart's domain.
I saw your eyes;
now castles rise in Spain!

and confesses his puzzlement:

> the only line that contains no rhyme is "A house in Iceland."
> This fact seemed curious to me since Hart was fond of rhym-
> ing and the "house" might have been anywhere. *Except,* the
> idea is: "Was my heart's domain. (It was cold.) Afterward,
> Hart concludes the idea: "I saw your eyes;/Now castles rise in
> Spain!"[13]

Yet suspecting Hart of *not* rhyming is as dangerous as suspecting
Shakespeare of not punning: in the very lines Engel quotes there
is a slant rhyme on *Ice*land and *eyes* and *rise,* then, I believe, a
faint tie rhyme: *I saw.*

Such unobtrusive rhyming may have prompted Hart to cite
"My Heart Stood Still" and its casually colloquial "I took one look
at you—that's all I meant to do" (instead of "I threw the book at
you") to defend himself against the charge of overrhyming. Still,
in the climactic line of the song—"un*til* the *thrill* of that moment
when my heart stood *still*"—he slips in a patented internal
rhyme. Hart's struggle to contain rhyme within the limits of conversational speech must have been fueled by a reviewer of *Connecticut Yankee* who invoked the vernacular spirits of Twain and
Mencken and cautioned:

> Occasionally the collaborators seemed to remember that they
> had been identified as modern Savoyards, and this seemed
> to please the attendant representatives of Gilbert and Sulli
> van, who were respectively Alexander Woollcott and Frank
> Sullivan. But if they want my advice, it's "Be yourselves,
> kiddoes."[14]

Hart seemed to be taking that slangy advice in the half-
dozen shows he and Rodgers turned out in the late 1920s. While
his addiction to rhyme could still produce such Gilbertian bend-
ers as "willing ham for Dillingham" and "Jeritza will shiver when
it hits 'er," he turned more and more to simple colloquial terms.
In *Present Arms* (1928), he built his lyric around the catch-phrase

"You Took Advantage of Me," but cleverly turned it from a belated accusation to a wry invitation:

> I'm so hot and bothered that I don't know
> my elbow from my ear.
> I suffer something awful each time you go
> and much worse when you're near.

The kn*ow*/el*bow*/*go* rhyme is there, accented lightly by the musical break, but it never disrupts the ebulliently conversational phrasing:

> So lock the doors and call me yours,
> 'cause you took advantage of me

Working from vernacular idioms, Hart adroitly ties together their imbedded metaphors—"hot and bothered," "cooked my goose" and "Here am I with all my bridges burned." At certain points he doubles the metaphoric knot with rhyme, letting "apple on a bough" grow out of "sentimental sap" to create a faint reversed rhyme: "sentimen*tal sap*" inverted by "*apple.*"

"You Took Advantage of Me" was also unusual in creating more of a sense of character than can be found in most love songs of the 1920s. Romantic lyrics aimed at Tin Pan Alley's popular market, as we have seen, were designed so that they could be sung by either a male or female performer—an important aspect of "standardization" in an industry increasingly geared, in the 1920s, to phonograph recordings and radio rather than sheet music. Even on Broadway, the songs written with an eye toward independent popularity were usually romantic duets; there might be "male" and "female" verses but the all-important chorus was "unisex"—sung in unison on stage and by either male or female singers on records. Even songs written as romantic solos on Broadway could go either way offstage. There were some exceptions, such as "The Man I Love," though as Ira Gershwin himself noted, the problem there wasn't in the title (which could be changed to "The Girl I Love" as "The Man That Got

Away" was later transformed to "The Gal That Got Away") but in the line "and he'll be big and strong."

The phrase "You took advantage of me," however, clearly implies a woman, but, as the lyric unfolds, that traditionally feminine accusation becomes a ploy that turns hapless victim into a coy seductress. The song *can* be sung by a man, of course, but then the rhetoric turns another way, as "he" slips into the role of the defenseless female and invites "her" irresistible assault. In "You Took Advantage of Me" Hart had found a wittily self-deprecating character who was perfectly suited to lyrics that were at once intricately rhymed yet casually conversational, and the song was the surprise hit of *Present Arms,* a "sassy and unregretful" number, Richard Rodgers recalled, that audiences liked far more than the traditional, "ardent" love songs.[15] Occasionally, however, Hart gave in to the sonority of Rodgers' music and strove for ardent, "poetic" heights. "With a Song in My Heart" (1929), for example, reveals how sappily sentimental Hart could be when he dropped his mordant guard:

> At the sound of your voice
> heaven opens its portals to me.
> Can I help but rejoice
> that a song such as ours came to be?

Significantly, when Hart reached for such melodramatic heights, he opted for the simplest of rhymes, reserving intricate ones for his wittily antiromantic songs.

By 1930, however, those clever rhymes began marking a different romantic motif—not ecstatic celebrations but masochistic addiction. The first lyric to register the self-lacerating use of rhyme was "Ten Cents a Dance" (1930), which took the character glimpsed in "You Took Advantage of Me" and fleshed her out with rough-edged slang and painful rhymes. It is ironic that a song so steeped in character and dramatic situation came from one of the most "unintegrated" musicals of the era. The show, *Simple Simon,* was a series of fairy-tale sketches, loosely held together by a dream plot, designed purely as a vehicle for slapstick comedian Ed Wynn. Just before it opened in Boston, the producer, Flo Ziegfeld, insisted Rodgers and Hart add an extra song for his female lead, Lee Morse. The team worked for one afternoon and came up with "Ten Cents a Dance," an extreme instance of the "interpolated" songs of musicals during the golden age—completely out of place within the dramatic context, geared purely for Tin Pan Alley's pop market. Its status became even more precarious the next day when Morse left the show. If Ruth Etting had not stepped in to play the role of the brassy taxi dancer, "Ten Cents a Dance" would have been cut. Etting's tough-little-girl delivery made the song not only the hit of the show, but one of the biggest hits in the early months of the Depression.

While Alec Wilder dismisses it musically as "at best, simply another competent pop tune," Rodgers himself praised his partner's lyric as a major transformation of the "torch song" genre of "Bill" and "Mon Homme." In expanding that genre, Hart created a pathetic but unpitying character he would rework again and again during the decade, his own transformation of the wittily self-deprecating lover in McCarthy's "You Made Me Love You" and "I'm Always Chasing Rainbows."

In such comic laments Hart found a new use for rhyme, a use that Anthony Burgess feels is peculiar to American popular song:

> The complex rhyme did not, of course, begin with Lorenz Hart. We find it in Robert Browning, who was a musician, and W. S. Gilbert, who wrote words for music, and also in Thomas Hood and in Barham's *Ingoldsby Legends,* but the Victorian intention was more heavily comic than witty: the grotesque rhyme was a signal that the rules of seriousness were being broken. With Hart and his followers there is a certain ambiguity of feeling, wit in the service of frustration or neurosis.[16]

The self-flagellating rhymes emerge in "Ten Cents a Dance" where, right from the opening of the verse, Hart takes the sweet rhymes of "Blue Room" and gives them a frustrated twist:

> I work at the Palace Ballroom,
> but gee, that Palace is cheap;
> when I get back to my chilly hall room,
> I'm much too tired to sleep

The "ball room"/"hall room" rhymes that obtruded into "Blue Room" are now buried under the relentless tread of assonance that matches the musical oscillations between two notes—G and E. The foot-dragging continues throughout the verse with "I'm one of those lady teachers—a beautiful hostess you know," until the last line, "The kind the Palace features at exactly a dime a throw" reverberates drearily with long vowels.

The same *i*, *e*, and *o* sound carry over into the patter section to underscore the singer's mechanical routine:

> Though I've a chorus of elderly beaux,
> stockings are porous with holes at the toes.
> I'm here till closing time.
> Dance and be merry, it's only a dime.

Despite the ten-cent economics, the singer's plight is one of imaginative enslavement: she is imprisoned in the "palace" because she can only see herself, romantically, as a beautiful princess waiting for Prince Charming:

> Sometimes I think I've found my hero,
> but it's a queer ro-
> mance

Split torturously over the monotonous *e*/*g* musical interval, "queer romance" rhymes back on "hero" and forward on "dance" as well as on the "*pansies*" who frustrate her romantic illusions.

Although Etting's delivery brought out the treadmill quality of music and lyric, she altered the words at a crucial point and lost not only a subtle rhyme but a link in the metaphoric pattern. Where Hart wrote,

> Trumpets are tearing my eardrums,
> customers crush my toes

Etting substituted "breaking" for "tearing" and thus cut his off-rhyme (*tearing/eardrums*), blurring the play of consonance and rhyme in and out of "customers," "crush", "trumpets," and "eardrums." Her substitution also disrupts the sequence of "tearing" imagery—the "tough guys who tear my gown," the "porous" stockings, the "tailors," "butchers," and "barbers" who all, in a lacerating pun, "rent me."

"Ten Cents a Dance" strikes a masochistic chord that reverberates through many of Hart's great lyrics of the 1930s, even the most winsome. "Dancing on the Ceiling" (1930), also uses the images of dance as torment: a surreal dancer on the ceiling turns the lover's bedroom into a chamber of delightful tortures:

> I try to hide in vain,
> underneath my counterpane;
> there's my love
> up above.

At Ziegfeld's insistence, "Dancing on the Ceiling" was cut from *Simple Simon*, though what Ziegfeld probably objected to was less the torment than the innuendo of "I whisper, 'Go away, my lover, it's not fair,' but I'm so grateful to discover he's still there."

For the next few years, Hart had to suppress both masochism and innuendo as he and Rodgers followed the Depression exodus of songwriters from Broadway to Hollywood. Hollywood found his lyrics "too flip," and he had to water them down to placate the producers. Rodgers called the Hollywood captivity "the most unproductive period of my life"; Hart, friends recalled, partied, drank, and spent heavily. Their best film, *Love Me Tonight* (1932), allowed Hart to indulge his antiromantic streak briefly in "Isn't It Romantic," which starts out quite sardonically as Maurice Chevalier, playing a Parisian tailor, muses about the prospects of married life:

> Isn't it romantic?
> While I sit around,
> my love can scrub the floor.

> She'll kiss me ev'ry hour
> or she'll get the sack,
> and when I take a shower
> she can scrub my back.

His song is picked up, first by a customer in his shop, then by a taxi driver, and, as the camera follows it out of the city into the French countryside, by a musician on a train, a platoon of soldiers, and a band of gypsies, until it finally reaches a princess, Jeanette MacDonald, who hears it from her moonlit balcony. When she takes up the song, however, the lyric is utterly transformed, decked out with poeticisms like "perchance," sentimental clichés like "secret charm" and "heart's command," and exclamations right out of operetta:

> Isn't it romantic
> merely to be young
> on such a night as this?

While cinematically innovative, the progression of "Isn't It Romantic" is lyrically regressive, forcing Hart from urban (and urbane) sophistication toward the elevated sentimentality appropriate for romantic castles in the countryside.

The more Hollywood made Hart strain for melodramatic effects, the more he sounded like a parody of himself:

> Lover, when I'm near you
> and I hear you
> speak my name
> softly, in my ear you
> breathe a flame

Here the enjambed "ear you" rhyme only points up the hyperbolic silliness of the image of fire-breathing passion.

Perhaps some of Hart's lyrical strain reflects the fact that he and Rodgers, working with directors like Rouben Mamoulian, tried to integrate songs into dramatic situations. Moving away from the notion of film musical as spectacle, where characters had to have an excuse for "performing" a song, Rodgers and

Hart wrote songs that emerged from character and plot, and they even experimented with rhythmic and rhymed dialogue. According to Rodgers himself, however, their timing was all wrong. When they arrived in Hollywood the first wave of musicals, brought on by sound films, had subsided. Then in 1933, just as they were experimenting with "integrated" songs and dialogue, came *Forty-Second Street,* with a renewed emphasis on song as spectacle—spectacle enhanced by the choreography of Busby Berkeley, who, ironically, had worked with Rodgers and Hart on Broadway in the 1920s.

For Hart, the final years of his Hollywood stint largely consisted of cranking out melodramatic lyrics, from "You Are Too Beautiful" (1933), with its clunking follow-up, "And I am a fool for beauty," to his last—and most loathed—assignment: putting English lyrics to Lehár melodies for *The Merry Widow.* Even Rodgers was amazed that his partner could turn out a lyric with such high-pitched sonorities as "Vilia, oh Vilia, don't leave me alone! Love calls to love and my heart is your own!" Hart's lowest moment in Hollywood, ironically, culminated in his most popular song. It started out as a song called "Prayer" for Jean Harlow in *Hollywood Party,* but when it was cut from the picture, Hart rewrote it as the title song for *Manhattan Melodrama,* and when it was cut again, he rewrote it as "The Bad in Every Man." Finally this lyrical tar baby received its final rebirth when music publisher Jack Robbins dared Hart to rewrite it, not as a theater song, or even as a Hollywood song, but as a straight Tin Pan Alley song. "Give me something us common people will understand," Robbins teased, "Y'know, a love song with June and moon and spoon." "How about 'Blue Moon,'" Hart shot back, then cynically strung together the very sort of rhymes he had mocked years before in his first meeting with Richard Rodgers:

> Blue moon,
> you saw me standing alone,
> without a dream in my heart,
> without a love of my own.

By 1935, Rodgers, who had been chafing in Hollywood's golden chains for years, had had enough. When Billy Rose invited him and Hart to write songs for his circus musical, *Jumbo* (staged, complete with elephants, at the Hippodrome), the team returned to Broadway, though not without some trepidation at a New York critic's musing, "What ever happened to Rodgers and Hart?"[17] They needn't have worried. Over the next few years, they wrote songs for such musicals as *On Your Toes* (1936), *Babes in Arms* (1937), and *The Boys from Syracuse* (1938), songs that vied with those of Cole Porter and the Gershwins to mark the high point of the golden age. Just before the advent of the "integrated" musical with *Oklahoma!*, they filled shows and films with songs that detached themselves, one after another, to become hits, songs that have endured as the "standards" we most often associate with these writers, songs where each lyricist give his unique stamp to saying "I Love You" in thirty-two bars.

Hart's characteristic stance in these lyrics is a tenderly masochistic one, underscored by subtly oppressive rhymes that unfold from the most ordinary turns of phrase. "Little Girl Blue," for example, is a more sophisticated version of Hart's toughnosed taxi dancer, and her rhymes are quieter but just as self-lacerating. In a line like "No use, old girl," Hart uses Rodgers' octave drop in "No use" to emphasize "u" enough so that it rhymes softly, but insistently, with "blue." In between those rhymes he deftly strings a sequence of double rhymes:

> you may as well surrender,
> your hope is getting slender.
> Why won't somebody send a
> tender
> blue boy
> to cheer a
> little girl blue?

As the double rhymes come more quickly—"send a/tender"—they underscore the motif of passing time which runs throughout the song. The singer, shrewdly aware of both her growing age and her youthful longings, sidesteps self-pity by addressing herself as "old girl" even as she portrays herself in the third-person as "little girl blue." She further undercuts sentimentality by scolding her child-ish self-caricature: "Sit there and count your little fingers" with its tiny "*sit/litt*le" rhyme. She even gets away with the hackneyed metaphor of teardrops and raindrops (and another *it* rhyme):

> Sit there and count the raindrops
> falling on you.
> *It*'s time you knew
> all you can count on is the raindrops

With a buried rhyme on "falling" and, on the melody's lowest note, "all," Hart deepens his theme with the same light touch as does the pun on "count."

Although his favorite lyrical setting remained that of the broken-hearted lover who "took it on the chin" but is defiantly "Glad to Be Unhappy," Hart could give a harsh curve to the sweetest of pitches. A protestation of love could sound like a street-taunt:

> My romance
> doesn't need a thing but you.

In its list of "I don't need's" Hart could allude back to his own romantic "castle rising in Spain" from "My Heart Stood Still." Similarly, he could celebrate "The Most Beautiful Girl in the World" with an antiromantic chip on his shoulder: she "isn't Garbo, isn't Dietrich!" Instead, he makes her sound like a nui-sance who

> picks my ties out,
> eats my candy,
> drinks my brandy

To top off his prosaic list of her "beauties," he taunts his listener's conventions just as Shakespeare did when he bragged about his dark lady's wiry tresses: "She'd shine anywhere—and she hasn't got platinum hair!"

In *Babes in Arms* (1937), Hart created a list song for just such a lady—a "tramp" from a ruggedly pastoral "Hobohemia" who pugnaciously defies the unconventional conventions of urbanity:

> I don't like crap games
> with barons and earls.
> Won't go to Harlem
> in ermine and pearls

An equally defiant character emerged from the tour-de-force, "Johnny One-Note," his primitive cry fitted to Rodgers' incessant device of pulling the melody line down to middle C. Hart enlivens the ear-shattering monotony with complex double and triple rhymes: "*gusto*" and "*just* overlorded the place," "yelled w*illy-nilly* un*til he* was blue in the face," and "got *in Aida indeed a* great chance to be *free.*"

From such thumping rhymes, he could shift to unobtrusive ones that underscore an affectionately mocking compliment:

> Is your
> figure
> less than Greek?
> Is your mouth a little weak?

"My Funny Valentine" nearly turns compliment to insult, with the off-rhyme "Is your/figure" accenting the skewed flattery. Similarly, the lopsided rhyme of "Your looks are laughable" with the six-syllabled "Unphotographable" further accents the disproportions of the "slightly dopey" beloved. Such imbalance shapes the entire song, the verse full of poetic archaisms such as "behold," "doth," and "knowest," the chorus laced with such colloquial idioms as "you're my fav'rite work of art."

Even love at first sight can bring the frustration of *déjà vu* (or the alcoholic's groping efforts to recall a black-out):

> It seems we stood and talked like this before,
> we looked at each other in the same way then,
> but I can't remember where or when

Although Wilder faults Rodgers' music in "Where or When" for its "plethora" of repeated notes, the monotony heightens the nagging torment. Hart's rhymes have the teasing simplicity of *déjà vu:* "then" and "when" echoing under "hap*pen*ing a*gain;* "where" picking up the last syllable of "remem*ber*" and resounding in "you're *wearing*"; then "wore" rhymed with three repetitions of "before" as the melody moves up in three interval steps with the climactic rhyme falling inconspicuously on "or."

While he could turn any romantic situation unpleasant, Hart's wittiest song from *Babes in Arms,* "I Wish I Were in Love Again," rang changes in his favorite figure of the jilted and jaded lover:

> The furtive sigh,
> the blackened eye,
> the words "I'll love you till the day I die,"
> the self-deception that believes the lie—
> I wish I were in love again.

The telegraphic list of phrases thrashes romanticism by punctuating a clichéd sequence of rhymes—"sigh," "eye," "die"—with "lie," but when Hart's world-weary lover strings out the rhyme with "I wish I were in love again," he confesses that, even knowing that love is all sham and cruelty, all pain and theatrical posturing, he nevertheless *still* longs to be "punch-drunk" and "ga-ga!"

The Porterish list of images grows increasingly violent, from "the conversation with the flying plates" to "the pulled-out fur of cat and cur" until even "broken dates" sounds destructive. The needling rhymes—"I *miss* the *kiss*es and I m*iss* the bites" and "*I* don't *like quiet*"—accent the abrupt phrases and keep their percussive bite even when the syntax grows more extended:

> when love congeals
> it soon reveals
> the faint aroma of performing seals,
> the double-crossing of a pair of heels

Still to want love—knowing how bad it feels, looks, and even
smells—is yet another of Hart's backhanded tributes to its power,
love seen not through the bright eyes of Romeo and Juliet but
through the baggy lids of Antony and Cleopatra.

Having pushed romance to one masochistic extreme, Hart
could vary that characteristic reversal the next year for *The Boys
from Syracuse* (1938):

> This can't be love because I feel so well—
> no sobs, no sorrows, no sighs!

Once again not only the diction but the telegraphic phrasing is
thoroughly conversational, punctuated by the simple *so/no/sorrow*
rhymes and an unrhymed allusion to one of his own earlier titles:

> My heart does not stand still—
> just hear it beat!
> This is too sweet
> to be love

Paradoxically, it is this skeptical sophisticate who gives love a
fresher twist than the ga-ga innocent of the 1920s "My Heart
Stood Still."

Just as Hart could reverse his usual masochistic approach to
love, he could also turn his usual technique inside out—creating
in "You Are So Fair" a tour de force of *not* rhyming. In place of
his dazzling double and triple rhymes, he weaves an antiromantic
lament that turns on repetitions of the same word, "fair," but
each time with a different meaning: from "you are so fair" (beautiful) to "you're not quite fair" (just), then "love affair," "bill of
fare," "See how you'll fare," and culminating in the wryly critical
"When I come to think it over, you're only fair" (average).

Hart had always been a difficult partner, but while they were
writing the *The Boys from Syracuse,* things got to the point where,
according to Rodgers, it was impossible to find him. Hart's alcoholism compounded another long-standing problem: while he
was a lyricist who wrote with incredible speed (a song like "The

Lady Is a Tramp," for example, was finished in a day), he could usually only work with the finished music before him. It was an enormous relief for Rodgers to work with Oscar Hammerstein, since Hammerstein was not only thoroughly stable but wrote lyrics first and gave them to the composer to set.

Strained as their partnership was, Rodgers and Hart continued to collaborate, on *Too Many Girls* in 1939, which produced a hit whose title reflected Hart's own worsening condition—"I Didn't Know What Time It Was." Then in 1940, for *Higher and Higher,* he found another alcoholic catch-phrase to twist into a romantic lament, "It Never Entered My Mind":

> Once you told me I was mistaken,
> that I'd awaken with the sun,
> and order orange juice for one

Hart's jilted lover is bemused by her own heartache and understatedly characterizes it as mild discomfort—"uneasy in my easy chair"—reflected verbally in uncomfortably rasping rhymes of "*or*der *or*ange juice *for* one." Even such languorous suffering has an erotic touch, from vaguely Freudian observations,

> you have what I lack myself

to wryly sensuous imagery,

> and now I even have to scratch my back myself

With clinical detachment, Hart's forlorn singer manages to turn heartache into an interesting, even refreshing, experience that has her singing "the maiden's pray'r again" and innocently wishing "that you were there again—to get into my hair again."

In the 1938 production *I Married an Angel,* Hart found his "ideal performer" in Vivienne Segal, the incarnation of the cynical sophisticate he had been writing for. "She says it and sings it like a lady," Hart would brag of her, "but with a twinkle in her eye."[18] One of the songs he wrote for her, in fact, was "A Twinkle in Your Eye," which she sang in the role of a jaded widow giving a coy lesson in corruption to an angel. He wrote for Segal again in *Pal Joey* (1940) and gave her his last great hit, "Bewitched." The show itself, based on John O'Hara's tale of double-crossing heels, appealed to Hart, and was as close as he would come to an integrated musical. Even though it became independently popular, "Bewitched" is a song that is intimately tied to a specific character and dramatic situation.

Sung by Segal as the hardened, horny, and well-heeled socialite Vera Simpson, the verse begins when she awakens after a drunken romp with Joey, her gigolo, who lies in a stupor beside her:

> After one whole quart of brandy,
> like a daisy I awake.
> With no Bromo seltzer handy,
> I don't even shake.

The jaded sophisticate who has "seen a lot—I mean a lot" now feels "like sweet seventeen a lot," yet even as she delights in it, she mocks her own recovered innocence. Relentlessly she keeps the infatuation purely sexual, vowing to "worship the trousers that cling" to a lover who is, "horizontally speaking," at "his very best." In lyrics that reviewer Brooks Atkinson found "scabrous," Hart had this weariest of all worldly lovers revel not in Joey's but in her own sexuality, narcissistically relishing being "dumb again and numb again, a rich, ready, ripe little plum again."

Yet just as innocence is so deliciously recovered, she takes a sobering look at her "half-pint imitation" of a lover and finds that romance as well as lust is a sham:

> Wise at last,
> my eyes at last
> are cutting you down to your size at last—
> bewitched, bothered, and bewildered no more.

A cynicism that triumphs not only over romanticism but over sexual ardor finally finds it is the *loss* of feeling that feels good:

> Couldn't eat—was dyspeptic,
> life was so hard to bear;

>   now my heart's antiseptic
>   since you moved out of there

Hart underscores this passage from sexual heat to clinical cool by using such passionate clichés as "again!" and "at last!" to celebrate the loss, rather than the discovery, of love. For his climax, he turns a French lament into a joyous cry over the conquest of sexual desire:

>   Romance—finis!
>   Your chance—finis!
>   Those ants that invaded my pants—finis!

Hart's own swan song, "Bewitched" was the last conceivable turn one could give to the theme of lost love, and it became independently popular only when lines like "worship the trousers that cling to him" were bowdlerized to "long for the day when I cling to him." Even today the final lines are cut by singers who want to make at least sex, if not love, sound appealing.

It is ironic that Rodgers and Hart reached the height of their fame just as their partnership, and the age it had inaugurated, was dissolving. In 1938 they were featured in a *New Yorker* "Profile" and even appeared on the cover of *Time*. A year later an article in the *Atlantic City Press* (July 9, 1939), under the headline "The Sophisticates," noted that "before 1925 you were content to express your romantic feeling" in the sentimental terms of "songs like 'I'll Be Loving You, Always,' 'Why Do I Love You?' and 'My Blue Heaven.' " But, the writer observed, "the sophisticated boys changed all that." With these "boys"—not just Rodgers and Hart but the Gershwins and Cole Porter as well— "popular songs took a definite turn for the less sentimental" and "with it popular song language changed from sweet and simple to slick and sophisticated." Under a picture of Rodgers and Hart, the story concluded that "with their appearance in Tin Pan Alley the popular song grew up."

It was a short-lived maturity for Rodgers and Hart—and for popular song. Although they worked together again, on *By*

Jupiter, and, just before Hart's death in 1943, on a revival of *Connecticut Yankee,* the partnership never recovered from Hart's refusal, in 1942, to work with Rodgers on a musical based on *Green Grow the Lilacs.* The play had been successful on Broadway, but Hart found its regionalism corny and was sure it could never make a successful musical. Rodgers then turned to Oscar Hammerstein, who had long been trying to get Jerome Kern to work with him on *Green Grow the Lilacs.* What Rodgers and Hammerstein both sensed was a shift away from cosmopolitan elegance toward regional simplicity—a shift already evident in John Steinbeck's novels, Thomas Hart Benton's paintings, and Aaron Copland's ballets (the Pulitzer Prize for drama in 1938 had, after all, gone to *Our Town*). When *Oklahoma!* opened in 1943, a new songwriting team was established, the Broadway musical became an "integrated" blend of songs and story, and an age of elegance, sophistication, and urbanity in popular song was over. It seems only fitting that an era that began with the cosmopolitan sparkle of "Manhattan" should close with the homespun yawp of "Oklahoma."

'S Wonderful: Ira Gershwin

What George was to music, Ira was to words.

HOWARD DIETZ

George S. Kaufman once listened impatiently as Ira and George Gershwin played one of their new songs; after fidgeting through references to Henry Ford, Robert Fulton, Thomas Edison, and Eli Whitney, the frustrated Kaufman interrupted at the line "They told Marconi wireless was a phoney" and demanded to know if this was going to be a love song. When the brothers finally got to "They laughed at me wanting you," Kaufman shook his head resignedly and said: "Oh, well . . ."

While Lorenz Hart's lyrics explore romance from every angle, Ira Gershwin seems to write about love only because he has to, or, rather, because he had to write about *something*. Though he once joked that without love "I'd be out of business," what really interested him was less the romantic message than the medium of language itself—the vocabulary, idioms, and phrasing of American speech. A romantic quarrel, for example, served merely as a clothesline where he could pair off English and American pronunciations of *either, neither, potato, tomato, after,* and *laughter*. Where Hart rang most of his changes on the broken-hearted lament, Gershwin situated most of his songs at the moment of falling in love, and his frequent references to inventors and discoverers—from Byrd to Dali—reflected not only the inspiration of his newly smitten lovers but his own linguistic creativity.

It was a creativity, moreover, that seldom manifested itself in rhyme. Like writing about love, rhyming seems something Gershwin did because he had to:

> They laughed at me wanting you,
> said I was reaching for the moon;
> but, oh, you came through—
> now they'll have to change their tune.

What is striking here are not the rhymes—indeed, they are the oldest *too*-for-*tee*'s in the book (though some of the *ee*'s are cleverly couched in *me* and *reaching*)—but the effortless progression of colloquial phrases that delineates the mounting glee of a singer more delighted with his triumph over doubters than with his romantic conquest itself. By his own definition, writing lyrics was an art of fitting words "mosaically" to music, and Gershwin's most brilliant effects came from finding the ready-made slang phrase to fit every rhythmic turn of his brother's music.

So perfectly matched were their meters that it's surprising the Gershwins did not begin their major collaboration until 1924, years after they had been writing songs separately. George had established himself as an Alley composer in 1916 with "When You Want 'Em, You Can't Get 'Em, When You've Got 'Em, You Don't Want 'Em." Ira, however, aimlessly drifted from stints with a carnival to working in the family-owned bath-house. He had tried writing society verse and managed to place a few satirical pieces in Don Marquis' column in the New York *Sun,* as well as in Mencken's *Smart Set.* The latter brought a one-dollar check, and a far more valuable piece of advice from the English playwright Paul M. Potter, who told Gershwin to "learn especially 'your American slang.' "[1]

Although he had begun toying with lyrics to some of his brother's melodies in 1917, it wasn't until a year later that it occurred to him to try becoming, as he put it, "a writer of what they call 'lyrics.' " His notes from this first collaboration, "The

Real American Folk Song (Is a Rag)," show that Ira was heeding
Potter's advice. Indeed, the manuscript doesn't even look like
poetry (except maybe that of e. e. cummings); instead, it consists
of truncated and oddly spaced vernacular bits that mirror his
brother's abrupt musical phrases:

> You jazz it
> As it
> Makes you hum.

When he and George sat down at the piano, moreover, Ira
dropped many of his rhymes and shortened the few long lines he
had written; "There's a happy, snappy, don't care a rappy sort of
I don't know what to call it," for example, was revised to an
unrhymed juxtaposition of slang phrases: "They lack a something: a certain snap." Just as George Gershwin had taken his
musical cue from Irving Berlin's "Alexander's Ragtime Band,"
Ira seems to have followed Berlin's lead in ragging vernacular
phrases against musical ones.

"The Real American Folk Song" was far from a real success,
however, and the brothers collaborated only sporadically over the
next six years. In 1919 George Gershwin wrote his first full-scale
show, *La-La Lucille,* with lyrics by Arthur Jackson and Buddy
DeSylva; the same year saw the biggest hit he would ever write,
"Swanee," with lyrics by Irving Caesar. Ira, meanwhile, collaborated with Vincent Youmans, Raymond Hubbell, and other composers, including an occasional song with George again, but he
insisted upon writing under the pseudonym Arthur Francis (derived from the names of another brother, Arthur, and his sister
Frances) so as not to ride on George's growing fame. It was only in
1924, after George's triumph with "Rhapsody in Blue" (whose
title had been Ira's suggestion—borrowed from the Whistler paintings he admired), that the two brothers settled into permanent
collaboration and "Arthur Francis" disappeared.

Lady, Be Good! starred the dance team of Fred and Adele
Astaire, and their big hit was "Fascinating Rhythm." "This
song's departure from the past is so marked," James Morris

observes, "that the listener is at a loss for a simile." Yet implicit
in Morris' description of the melody as "an intricate accumula-
tion of bits and pieces, fragments stitched together"[2] is the metaphor of collage, the Cubist updating of the mosaic figure that
Ira Gershwin himself invoked to describe the art of lyric writing. The metaphor is certainly apt for George, who had already
begun collecting modern art and aspired to "write the way
Rouault paints."

Yet Ira showed himself equally adept at fitting verbal shards
to the nervous musical accents, framing internal *is* rhymes between harshly alliterating *d*'s and *t*'s, then matching the frenetic
dotted-eighth/sixteenth-note pattern with irritated contractions:

> So darn persistent,
> the day isn't distant,
> when it'll drive me insane

Like Irving Berlin, Ira had found that contractions'll not only
keep a lyric colloquial, they'll intensify the musical pace. But
things got even slangier—and faster—if words were dropped
altogether:

> —comes in the morning
> without any warning

When his brother suddenly slows down—without any warning—
Ira's lyric becomes a study in elongation with "and hangs a-rou-nd
all day," where *round* is stretched over two notes.

All of this ragging of word against music was still only in
the verse. When they got to the chorus, where the brief rhythmic theme repeated itself over and over for eight bars—and
each time on a new accent—"there was," George recalled,
"many a hot argument between us" over "where the accent
should fall in the rest of the words."[3] Ira had wanted to emphasize the shifting rhythm with simple masculine rhymes on a
single accented syllable, but when George demanded a two-
syllable feminine rhyme (where the second syllable is unaccented), Ira came up with a Hart-like juxtaposition of an archaic

metaphor—"I'm all a-quiver"—with a modern one—"I'm always shaking just like a flivver" (an apt enough simile for the flivver-driven 1920s).

Equally artful was the way Ira matched his brother's every rhythmic move with a perfectly colloquial catch-phrase: "You've got me on the go," "Won't you take a day off?," and "Won't you stop picking on me?" Perhaps the cleverest device in the song is a ragged mismatch of words and music that emphasizes repetitiveness without itself becoming monotonous: as Gerald Mast notes, when Ira repeats his title phrase, he stretches "Fascinating" over "the boundary between measures two and three" so that George's "downbeat falls in the third measure without any new word to announce it—a downbeat both felt and missed. This tiny trick provides the fascination of this fascinating rhythm, both for the singer, possessed by it, and the listener, who soon will be."[4] Like Gus Kahn and other lyricists of the 1920s, Ira Gershwin let such repetitive music generate his theme, but instead of writing about a prisoner of love trapped on the treadmill of fate, he transposed that victim into a "doped-out" addict, not of romance but of a rhythm as tormenting as one of Hart's lovers.

In the title song from *Lady, Be Good!*, Ira Gershwin wrote the first of numerous lyrics he later classified under such categories as "The Importunate Male," "The Importunate Female," "The Element of Time," "Findings," and "Euphoria"—all variations on his favorite theme of falling in love. It was a theme that gave free rein to his verbal inventiveness, and it was equally perfect for his brother's exuberant rhythms. In "Oh, Lady Be Good!," those rhythms pleaded for an importunate lyric with long whole- and half-notes interspersed with brisk triplets. Ten years earlier, in Kern's "They Didn't Believe Me," a triplet undid the lyric, but Ira relished the triplet as a way to fragment the first three syllables of "misunderstood." The triplet placed just enough emphasis on "mis" so that it rhymed with "I must win some winsome *miss*." Cleverer than the rhyme, however, is the play between shifting musical rhythms and lyrical contraction and elongation. Where one expects the contraction "I'm," Gershwin spells out "I am," but then crunches "awfully" into "so awf'ly," yet stretches those two syllables out over two long half-notes.

Probably the most famous song from *Lady, Be Good!*, the romantic ballad "The Man I Love," was cut from the show after its lukewarm reception in Philadelphia. In 1927 it was salvaged for *Strike Up the Band,* but when that show folded out of town, it was interpolated into *Rosalie,* only to be cut again before opening night. What Ira called "this thrice-orphaned song" first became popular as a nightclub song in London and finally made it as an independent hit in America through sheet music and record sales. Thus "The Man I Love" epitomizes the unintegrated character of theater songs of the 1920s, when a lyric had so little relation to plot or character it could be freely shunted from show to show. Such songs, as we have seen, were aimed at Tin Pan Alley's pop market, and it is not surprising that "The Man I Love," one of the Gershwin's biggest hits, should reach that market without even the mediation of a musical.

What is surprising is that, as Ira himself noted, it was one of the few songs whose lyric violated the Alley axiom that songs must be equally performable by male or female singers. To capture a girlishly hesitant—but determined—longing, Ira relies on parallel, appositional phrases that keep stopping—but then relentlessly drive forward with the music:

> Maybe I shall meet him Sunday . . .
> maybe Monday . . .
> maybe not . . .
> still I'm sure to meet him one day . . .

The clichéd daydreams of "a little home just meant for two; from which I'll never roam" are enlivened with down-to-earth asides— "Who would? Would you?"—and such frank acknowledgments as "And though it seems absurd, I know we both won't say a word," with its grammatically skewed "we both won't."

Lady, Be Good! launched the Gershwins on a series of successful shows that, like most musicals of the 1920s, were loosely struc-

tured affairs designed to showcase songs. Almost all of Ira's hits
in these years were songs that rang changes on his favorite romantic situation of falling in—rather than out of—love. In 1925
he homed in on that euphoric moment for "That Certain Feeling," the big hit from *Tip-Toes*. Using throwaway rhymes, Gershwin characteristically puts all his emphasis on matching his
brother's syncopation with a lyric so "ragged" many singers
couldn't follow it. It was the first of many feuds Ira would have
with singers who missed his "rhythmic point." In *Lyrics on Several
Occasions,* he spelled out his verbal syncopation on paper, show-
ing how he wanted two syllables collapsed onto one note after a
one-beat rest:

> That certain feeling!
> (Beat) *Thefirst* time I met you!
> I hit the ceiling!
> (Beat) *Icould* not forget you!

Combined with the telegraphic syntax, the rhythmic crunch
heightens the abruptness of the song and captures the gleeful
imbalance of love at first sight:

> Grew sort of dizzy!
> Thought, "Gee, who is he?"

That rhythmic imbalance, however, produces an odd balance of
its own by collapsing the six-syllable lines to match the five-
syllable ones.

It was during the run of *Tip-Toes* that Ira received a flatter-
ing letter from Lorenz Hart. Apologizing for not coming up after
the show to introduce himself and tell "you how much I liked the
lyrics," Hart ironically explained, "I had imbibed more cocktails
than is my wont." He then asserted that "your lyrics, however,
gave me as much pleasure as Mr. George Gershwin's music. . . .
It is a great pleasure to live at a time when light amusement in
this country is at last losing its brutally cretin aspect. Such delica-
cies as your jingles prove that songs can be both popular and
intelligent." Before closing, however, Hart took the "liberty of

saying that your rhymes show a healthy improvement over those in '*Lady, Be Good!*' "[5]

A connoisseur of rhyme such as Hart must have been even more delighted with the Gershwins' next hit, "Someone to Watch Over Me," (1926), since it had a cleverly split rhyme: "Although he may not be the kind of *man some* girls think of as *handsome.*" What makes the rhyme doubly effective is its spondaic rhythm, which slows down the lyric with the music. It may have been this double change of rhythm that inspired George Gershwin to try playing the whole song, which had started out as a "fast and jazzy" dance number, in a slower tempo. The experiment transformed "Someone to Watch Over Me" into a ballad, much like "The Man I Love," yet Ira's telegraphic syntax still gave the lyric a nervous impatience that matches the music:

> Lookin' ev'rywhere,
> haven't found him yet;
> he's the big affair
> I cannot forget—
> only man I ever think of with regret.

The hurriedness is heightened by clipping and contracting words, though again he varies the pace by spelling out "cannot." In the refrain, too, the contractions give what might have been a sentimental plea a slangy ease: not "there is" but "There's a somebody I'm longing to see." At the end, a contraction even gives the title a colloquial twist: "Someone *who'll* watch over me."

While "Someone to Watch Over Me" looks forward to the moment of falling in love, " 'S Wonderful" (1927) revels in it, pushing linguistic playfulness to new extremes to register that euphoria. Once again Ira complained of singers who miss the whole point of such playfulness when "they formalize the phrases to '*It's* wonderful,' '*It's* marvelous, '*It's* paradise.' " The lyric not only toys with vernacular diction and phrasing but it probes the very processes, such as slurring words and clipping suffixes, that create colloquial speech. By exaggerating these processes, the lyric makes language itself the wondrous subject, a

linguistic exuberance that reflects the joyous confusion of falling
in love.

   While working on *Lady, Be Good!*, Gershwin had heard one
of the comedians in the show, Walter Catlett, "clipping syllables."
In 1926 he had used the device sparingly when he put words to
Phil Charig's melody, "Sunny Disposish." But in " 'S Wonderful,"
the process devours the verse:

> don't mind telling you,
> in my humble fash,
> that you thrill me through
> with a tender pash.

Such snipping undercuts the sentimental hyperboles and refreshes the tritest of romantic rhymes:

> when you said you care,
> 'magine my emosh;
> I swore then and there,
> permanent devosh.

Even as it affirms passionate heights, the tone is casually understated, and in what Gershwin called the "vest" of a song—the two
lines that lead from the verse to the refrain—he uses slang to
dampen the ardor:

> You made all other boys seem blah;
> just you alone filled me with AAH!

As in the poetry of cummings, Gershwin enlivens a word by
switching it to another syntactic position, in this case using an
exclamation as a noun.

   In the refrain he reverses gears: instead of truncating suffixes, now he clips off the root of the contraction and lets the
dangling apostrophe *s* slide onto the next syllable:

> 'S awful nice! 'S paradise—
> 'S what I love to see!

In the release he changes pace again, using a long triple rhyme—
"glamorous"/"amorous"—for an extended musical phrase that
consists of little more than one note repeated for the entire eight
bars. Even those rhymes, however, serve to highlight his suffixes
and lead back to the "ous" of "marvelous."

In "How Long Has This Been Going On?" (1927) Gershwin
turned his favorite motif of first love yet a different way. Instead
of the gleeful exuberance of " 'S Wonderful," "How Long Has
This Been Going On?" responds to a blissful first kiss with irritation. The catch-phrase is something an irate mate would say to
an adulterous spouse (or an exasperated parent to a naughty
child), but here it's the lovers who scold themselves for discovering the joys of kissing so late:

> I could cry salty tears;
> where have I been all these years?

As George's music makes a sudden bluesy drop on "all," Ira
disrupts yet another catch-phrase—"where have *I* been all these
years"—and emphasizes the distortion by rhyming *all* with *salty*.

In this lyric Ira creates a wryly comic character, one who
laments "What a dunce I was before," then continues to be a
dunce by failing to come up with the proper definition:

> There were chills up my spine,
> and some thrills I can't define.

Finally the right definition comes in a simile even the classroom
dunce would know: "I know how Columbus felt finding another
world." Although he relies more on character and imagery here,
Ira still laces his lyric with slangy contractions: "Who'd 'a'
thought I'd be brought to a state that's so delirious?"

He used comic caricature to give first love yet another twist
in "I've Got a Crush on You" (1928), where a newly smitten
lover congratulates his beloved on *her* catch of *him:* "How glad
the many millions of Annabelles and Lillians would be—to capture me." Then, after praising her not for her beauty but for
her "persistence," he blithely adds insult to injury: "It's not that

you're attractive—but oh, my heart grew active when you came into view." In the chorus he completely forgets about his beloved and concentrates on his own newly discovered depths of feeling—"I never had the least notion that I could fall with so much emotion." Then he even magnanimously forgives his own sentimental indulgence: "The world will pardon my mush." Such unabashed narcissism undercuts sentimentality even as the language gets increasingly crushed and mushed—"could you coo?" oozing into "kootchy-koo." Ironically, a song the Gershwins wrote as a fast and unsentimental comedy duet became a hit only after Lee Wiley recorded it at a slow tempo, backed by syrupy organ music.

Gershwin's increasing interest in lyrics that reflected character and registered a comic situation signaled his shift, in the 1930s, to political satire in the operetta tradition. The last great frothy Broadway musical he wrote with George was *Girl Crazy* (1930), a show that produced not one but a string of hits on Tin Pan Alley. The rouser, "I Got Rhythm," gave free rein to Ira's vernacular play and made a star of Ethel Merman. Given its seemingly perfect fit of word to music, it is surprising to learn that Ira had extreme difficulty writing the lyric. The difficulty stemmed from the fact that each A section opened with the same repeated notes, and no matter what rhymes he used, Ira complained that "they seemed to give a pleasant and jingly Mother Goose quality to a tune which should throw its weight around more." Finally he struck upon the "daring" solution of abandoning rhyme; in its place he used alliteration (repeating *m*'s and harsh *k* and *g* sounds) that ran through a series of brusque slang phrases built around the ungrammatical device of substituting "got" for "have."

> I got rhythm,
> I got music,
> I got my man—
> who could ask
> for anything more?

Here Gershwin, like many modern American poets, used parallel colloquial phrases to give structure to unrhymed "free verse." The only rhymes in the chorus come in the release:

Old Man Trouble,
I don't mind him—
you won't find him
'round my door

Yet even here, consonance, telegraphic phrasing, and the clipped " 'round" keep up the conversational punch. He also used rhymes in the verse—and even proper grammar ("Look at what I've got") to set up his slangy chorus. Still, he complained, many singers missed the point by editing the chorus to "*I've* got rhythm, *I've* got music."

Just as "I Got Rhythm" made a star out of Ethel Merman, the romantic ballads of *Girl Crazy* highlighted another newcomer to Broadway, Ginger Rogers. "Embraceable You" was another of the Gershwins' chastely sensuous first-love songs. Although Ira uses a three-syllable rhyme (embraccable/irreplaceable/silk-and-laceable), it characteristically serves to highlight his play with suffixes: "-able" had already become the ad-man's all-purpose additive (even F. Scott Fitzgerald's teenagers were swooning over each other's "kissable" mouths). Ira heightens the grammatical play by using George's musical rests to separate the suffix from the root: by severing the -sy suffix in "tip/sy in me" and "gyp/sy in me," he captures the slightly discombobulated feel of love at first sight.

Ira could give even heartbreak the same aura of fresh discovery he used for falling in love. In another of Ginger Rogers' songs from *Girl Crazy*, "But Not for Me," he juxtaposes the stale clichés of melodrama, such as "Fate supplies a mate," against newly minted slang—"It's all bananas!"—to register the shock of finding that the romantic formulas don't come true in reality. He even capitalizes the clichés to stress the fact that he is quoting them:

With Love to Lead the Way,
I've found more Clouds of Gray

> than any Russian play
> could guarantee.

The theatrical metaphor runs through the lyric, turning it into a list song, composed not of Porterish images but of hackneyed formulas:

> I was a fool to fall
> and Get That Way;
> Heigh ho! Alas and al-
> so, Lackaday!

Not only does the conversational after thought "and also" provide a split rhyme, it also adds to the list-like monotony. The theatrical metaphor culminates in the last item in the catalog, "When every happy plot ends with the marriage knot," refreshed with a snappy pun on the title that allows him to indulge his penchant for double negatives: "and there's no knot for me."

Ira's love for contracting and stretching words got free rein in another hit from *Girl Crazy*, "Bidin' My Time," designed as a parody of Tin Pan Alley's "country" tunes and sung by a quartet of hillbillies. He recalled that, back in 1916, when he was still trying to write witty verse, he closed a quatrain with the ironical resolve:

> Some day when I'm feeling especially brave,
> I'm going to bide my time.

When he picked up the catch phrase for a song, however, he lazily clipped and elongated his words:

> I'm bidin' my time,
> 'cause that's the kinda guy I'm.

By dragging "time" over two notes, he highlights his theme of rural indolence, an indolence reflected in such locutions as "like that Winkle guy I'm" and the monotonous *i*-rhymes that finally don't even bother going beyond homonyms:

> chasin' way flies,
> how the day flies,
> bidin' my time

In the last line he even drops the "I'm," a lyrical truncation that accords with his own suggestion to his brother that the music be reduced from the traditional thirty-two bars to twenty-four.

Even before *Girl Crazy*, the Gershwins had begun moving away from such boy-meets-girl musicals toward political satire. Although it was hailed by critics as an "entirely new genre," however, *Strike Up the Band* (1927) closed after a few weeks in Philadelphia. Much of the failure has been attributed to the acerbic cynicism of George S. Kaufman's script, but audiences seem to have been equally put off by the Gershwin songs, which departed from the standard thirty-two-bar AABA pattern into what Ira called "patter songs, finalettos, recitatives, caroled couplets, and choral quatrains." These were not songs aimed at Tin Pan Alley but what he called "programmatic lyrics" that were inextricably tied to the libretto. In many of them, Ira was clearly emulating the witty patter of Gilbert and Sullivan rather than flashing his own colloquial style. So steeped was the show in the Savoyard tradition that for once Ira's words came before George's music. Nothing could save *Strike Up the Band*, however; on one of its last performances, Ira and Kaufman saw two elderly gentlemen, dressed in evening clothes, enter the theater. "Who are they?" Kaufman wondered aloud. "Oh, that's Gilbert and Sullivan," Gershwin replied, "coming to fix the show."

In 1930, however, *Strike Up the Band* was revived. Morrie Ryskind was brought in to tone down the political barbs in Kaufman's book, and a few romantic ballads were added. Even with songs like "Soon," however, Ira seemed to lose his deft handling of conversational speech, stumbling into such poetic inversions as "two hearts as one will be blended." Still, the revised version of *Strike Up the Band* was successful enough to prompt the Gershwins, Kaufman, and Ryskind to follow the course of political satire with *Of Thee I Sing* (1931). Here they moved even further toward comic opera—and away from popular song. Ira insisted that songs be integrated into the plot, and he bragged that the show contained no lowly Alley "verse-and-chorus songs":

> There is a sort of recitative running along and lots of finales and finalettos. It has meant easier work for both of us. It is hard to sit down and stretch out some idea for thirty-two measures. In this show you develop ideas, condensing pages of possible dialogue into a few lines of song. And George found it easier to write these measures, too, though he works much the same at anything he attempts.

While none of the songs became popular—indeed they were practically designed *not* to—the loss was more than offset for Ira, Kaufman, and Ryskind by the receipt of a Pulitzer Prize—the first ever for a Broadway musical (although George Gershwin, as composer, was deemed ineligible for the literary trophy).

Of Thee I Sing has been hailed, along with *Show Boat,* as one of our first integrated musicals—more than a decade ahead of *Oklahoma!*—but Broadway's gain was Tin Pan Alley's loss. Buoyed by the success of *Of Thee I Sing,* the Gershwins turned, in 1933, from writing popular songs to writing other political satires such as *Pardon My English* and *Let 'Em Eat Cake.* Neither of these turned out to be theatrical successes, however, nor did they produce independent hits, except for "Mine" and "Isn't It a Pity," with its insouciant, Porterish turns such as "My nights were sour—reading Schopenhauer" and "You—reading Heine; I—somewhere in China."

Thus the early 1930s were as dry a period for Gershwin popular songs—albeit for very different reasons—as they were for Rodgers and Hart. In 1935, the year that Rodgers and Hart gave up on Hollywood and returned to Broadway, the Gershwins undertook an even more ambitious movement—toward opera—with *Porgy and Bess.* George had been interested in turning Du-Bose Heyward's *Porgy* into an opera ever since he had read the novel one insomniac night in 1926. Negotiations with Heyward had been shelved, however, since Dorothy Heyward first wanted to adapt her husband's novel into a play. Plans for an opera fell through again in the early Depression years when a financially pinched Heyward negotiated with Al Jolson, who wanted to do *Porgy* himself (in blackface!), with songs by Hammerstein and Kern.

Although the Theater Guild finally managed to bring the Gershwins and Heyward together, *Porgy and Bess* was largely a collaboration between George Gershwin and DuBose Heyward. Most of the songs, such as "Summertime," were poems by Heyward that George Gershwin set to music, and that music moved still further from the thirty-two-bar AABA structure. Ira's major contribution was largely in the form of providing the more sophisticated lyrics for Sporting Life. Although his name was listed as co-lyricist, the only songs Ira Gershwin included when he compiled *Lyrics on Several Occasions* were "I Got Plenty o' Nuttin'," "It Ain't Necessarily So," and "There's a Boat Dat's Leavin' Soon for New York." The first two, as Sigmund Spaeth notes, were the show's only "frankly popular songs" that soon "developed independently into permanent hits."

Both, moreover, really were *popular* songs—where the music came first. "I Got Plenty o' Nuttin' " began when George played a melody for Ira and Heyward. In one of the few times a title came easily to him, Ira Gershwin recalls he quickly came up with the catch-phrase "I Got Plenty o' Nuttin' "—a spinoff from "I Got Rhythm"—then followed with "an' nuttin's plenty fo' me." At that point, Heyward asked to try his hand at writing lyrics, though it still took Ira's experienced ear to know "that many a poetic line can be unsingable, that many an ordinary line fitted into the proper musical phrase can sound like a million."[6]

But it took even the experienced lyricist a while to realize that his dummy title for "It Ain't Necessarily So" should be the real one. Like all dummy titles, the phrase was merely intended to help the lyricist remember the musical rhythm, in this case one where the accent fell on the "second, fifth, and eighth syllables." (Ira later observed that he might just as easily have used "*Tomorrow's* the *Fourth* of July" or "An *order* of *bacon* and *eggs*"). But the flippant sentiment of "It ain't necessarily so" perfectly suited a rhythm that turned out to be the same metrical pattern that governs the most irreverent of poetic forms—the limerick. Such a fascinating rhythm may well have inspired Sporting Life's sacrilegious sermon, with its winking references to Pharaoh's daughter (who "fished Moses—she *says*—from dat stream") and the plight of Methuselah ("who calls dat livin' when no gall'll give in to no man what's nine hundred years?"). Gershwin even added an "Encore Limerick":

> 'Way back in 5,000 B.C.
> Ole Adam an' Eve had to flee.
>> Sure, dey did dat deed in
>> De Garden of Eden—
> Buy why chasterize you an' me?

Here Sporting Life mocks the very confected black dialect he mouths, carrying over its "de's" to make even "Garden of Eden" sound like Uncle Remus talk. Like the stereotypical black preacher, too, he tosses highfalutin words like "chasterize" into his slangy pastiche. The mixture of high and low diction produces rhymes like "li'ble" and "Bible," "Goliath" and "dieth," and, for Jonah, "made his h*ome in* that fish's abd*omen.*" That very mixture was present in the fortuitous title-phrase with its counterpoint between an earthy "ain't" and the arch "necessarily."

Even with such comic songs, and the magnificent operatic numbers like "My Man's Gone Now" and "Bess, You Is My Woman Now," *Porgy and Bess* was neither a critical nor a popular success in 1935. Despite George Gershwin's efforts to keep it going after it closed in New York, the tour lasted only a couple of months. Exhausted and disappointed, he took a trip to Mexico, hoping to find new inspiration in Mexican music. Ira, however, went back to writing popular songs and had his first genuine hit in years. Teamed with composer Vernon Duke (who, at George Gershwin's suggestion, had changed his name from Vladimir Dukelsky), he wrote songs for the *Ziegfeld Follies of 1936*—an old-style revue, lavish and loosely structured, about as far as Broadway could get from an integrated musical. Such a revue called for free-standing songs, and in "I Can't Get Started," Ira recaptured his colloquial '20s style, infusing it with the sophisticated allusions Cole Porter had made a hallmark of popular songs in the 1930s. The singer—initially a man but, as the song grew in popularity, Ira had to write additional lyrics for female singers—has Porter's nonchalant sophistication, but the situation is still Gershwin's favorite:

> I'm a glum one; it's explainable,
> I met someone unattainable.
> Life's a bore,
> the world is my oyster no more.

Giving first love a rueful twist, he keeps sentimentality at bay by once again having the singer celebrate himself: "I've settled revolutions in Spain," "*Green Pastures* wanted me to play God," and "I'm written up in *Fortune* and *Time*." Such self-aggrandizement then turns to backhanded compliment, underscored by bemused irritation:

> The market trembles when I sell short;
> in England I'm presented at court.
> The Siamese Twins I've parted—
> but I can't get started with you.

Clever as the images are, the best touches in the lyric come when Gershwin doesn't try to out-list Porter but, characteristically, laces the singer's exasperation with bits of slang:

> I've got a house—a show-place—
> but I get no place with you

Some singers have made the lyric even more Gershwinesque with a double negative: "But I can't get no place with you." Gershwin sets up his slang with elegant poetic inversions,

> The upper crust I visit,

which are then debunked with direct speech:

> but, say—what *is* it—with you?

In the release, with its persistent return to the same whole note, the music itself can't get started, and Gershwin matches it with equally sputtering rhymes that interrupt, rather than advance, the syntax:

> You're so supreme—
> lyrics I write of you, scheme—
> just for a sight of you, dream—
> both day and night of you

All of these false starts then explode in the exasperated "And what good does it do?"—another instance of Ira Gershwin's knack of placing an utterly simple catch-phrase at an emotional climax, giving vernacular luster to both phrase and setting.

In the late summer of 1936, George Gershwin, too, returned to popular song—shelving plans for operatic and symphonic compositions to join Ira on a westward trek to Hollywood. It was their second venture in films; a brief stint in the early 1930s had produced *Delicious* and a film version of *Girl Crazy*. When they made their second trip, however, they did not have a string of Broadway successes behind them—in fact the studio bosses were concerned that, after *Porgy and Bess*, they were too "highbrow." George tried to assuage those concerns by telegramming ahead that the brothers Gershwin were once again out to "WRITE HITS!" Fortunately, RKO gave them a contract for the best Hollywood could offer songwriters—an Astaire-Rodgers film. The assignment gave the brothers some trepidation, for in this, the seventh Astaire-Rogers film, they would be following in the footsteps of Berlin and Kern.

But the Gershwins followed those acts with not one but two Astaire films in 1937, *Shall We Dance?* and *A Damsel in Distress*. The films made good on George's vow, producing an array of hits not seen since *Girl Crazy:* "They Can't Take That Away From Me," "Let's Call the Whole Thing Off," and "They All Laughed" from *Shall We Dance?;* "Nice Work If You Can Get It," "A Foggy Day," and "Things Are Looking Up" from *A Damsel in Distress*.

Before George Gershwin's death on July 11, 1937, they also completed "Love Walked In" and the chorus of "Love Is Here to Stay" for the 1938 *Goldwyn Follies*.

It's been said that "Hollywood killed George Gershwin," but in fact films revived the Gershwin brothers as popular songwriters. While many see these scores for films "as a regrettable if inevitable decline from a previous peak of innovative achievement and commercial disappointment"[7] on the musical stage, they represent one of the crowning achievements of the art of saying "I Love You" in thirty-two bars. However much one might lament the lost operettas, folk operas, political satires, and other integrated musicals that might have been, the songs the Gershwins wrote in their last year of collaboration are as much jewels of the golden age of popular song as are those that Rodgers and Hart wrote for Broadway during the same *annus mirabilis*.

With his blend of nonchalance and sophistication, Astaire was the perfect voice for Ira's urbane, conversational style. Astaire's easy grace even shaped the structure of the films, fulfilling what Rodgers and Hart had tried to create in film musicals— that songs, instead of being used for "spectacle," could emerge naturally, as if they were part of the dialogue. For example, "They Can't Take That Away From Me," from *Shall We Dance?*, begins with a verse that Astaire delivered more as conversation than song:

> Our romance won't end on a sorrowful note,
> though by tomorrow you're gone

One barely hears the light rhymes "Our ro"/"sorrow"/"though"/ "tomorrow," which work with the musical triplets to project the singer's nonchalant heartbreak. That chatty flow carries through the verse, which doesn't so much end as spill over syntactically into the refrain: "But though they take you from me, I'll still possess . . . the way you wear your hat."

"They Can't Take That Away From Me" was another of Ira's jaunty excursions into Hart's territory of romantic lament, but characteristically he takes the high road that makes the loss of

love almost as exuberant as its discovery. Where a lyricist like
Hammerstein or Howard Dietz would have strained for poetic
images "to remember you by," Gershwin opts for the utterly
prosaic: "the way you sip your tea . . . the way you hold your
knife . . . the way you sing off key." The lover who notes these
most trivial details, however, infuses them with understated passion, particularly when he abandons detail altogether with a casual flourish:

> The mem'ry of all *that*—
> No, no! They can't take that away from me!

The foregrounding, with rhyme and musical accent, of the insignificant linguistic bit *"that"*—*that* is Gershwin at his deftest: on
the one hand it seems to minimize "all *that*," yet, in this context, it
hints at unspoken depths of feeling in the "that" of the title
phrase.

In the release he engineers another structural shift that
keeps the lyric sounding more like conversational banter:

> We may never, never meet again
> on the bumpy road to love,
> still I'll always, always keep
> the mem'ry of

Just as the verse spilled over into the refrain, here the release
ends with a dangling "of," its suspended syntax bridging back
into the final A section with "the way you hold your knife."

Seamless as that lyrical web is, Ira could not resist reworking
it when he wrote *Lyrics on Several Occasions:*

> Studying this lyric, I see that the release could have been im-
> proved to:
>
> > We may never, never meet again
> >   On the bumpy road to love,
> > Still I don't know when
> >   I won't be thinking of—
> > The way you hold your knife . . .

> which would rhyme "again" and "when" on important notes
> where no rhyme now exists; also, the sequence "I don't know
> when I won't" contains a double negative, a form of phrasing I
> sometimes find myself favoring over a simple affirmative.[8]

But "half an hour later," he says, "no dice." In sticking with his
original, he characteristically sacrifices rhymes for a more prosaic
lyric and even gives up the gorgeously convoluted double negative: " 'always, always' following 'never, never,' plus all the 'No,
no!'s make a better-balanced refrain." Hollywood was, however,
still Hollywood: although "They Can't Take That Away From
Me" was nominated for the Academy award in 1937, it lost out to
"Sweet Leilani."

The Gershwins' other songs from *Shall We Dance?* and *A
Damsel in Distress* also sparkle with the same made-for-Astaire
ease. The latter film placed Astaire in London, and in "A Foggy
Day," Ira strung anapests together to create a fidgety monologue, "What to *do?*/What to *do?*/What to *do?*," then broke up his
meter for a prosaically resigned, "the outlook was decidedly
blue." But at his favorite moment of love-at-first-sight, Ira is all
iambs: "For *sud*/-den*ly*/I *saw*/you *there!*" That same euphoric situation got another colloquial formulation in "Things Are Looking
Up," with Ira's characteristic ragging of lyrical syntax against
musical phrasing:

> and it seems that suddenly I've
> become the happiest man alive

While Astaire could observe the subtly skewed phrasing here,
most singers, Ira again complained, tried to straighten it out by
singing

> and it seems that suddenly
> I've become the happiest man alive

No singer, however, missed the playful point of a Gershwin
lyric more than the London chorus girl who sang "You say
eyether and I say eyether, you say nyether and I say nyether."

"Let's Call the Whole Thing Off," from *Shall We Dance?*, a romantic parting that ends with a calling-off of the calling-off itself, is
typical of these last Gershwin songs—almost all are exuberant
and optimistic. Some, like "Nice Work If You Can Get It," from
Damsel in Distress, even make a cheerful nod toward the Depression, as does "They All Laughed," from *Shall We Dance?*. There,
as already noted, Ira had the tricky job of following his brother's
abrupt musical truncation of an expected seven-syllable line to
one:

> They all laughed at Rockefeller Center—
> now they're fighting to get in;
> they all laughed at Whitney and his cotton
> gin!

Not only is Ira's piece-work as inventive as the characters he
celebrates, his list of inventors and discoverers who overcome
failure and derision makes for a Depression-era anthem of faith
in American resiliency. Even the catch-phrase title is borrowed
from an advertisement slogan selling that most American of com-
modities, self-improvement: "They All Laughed When I Sat
Down to Play the Piano."

It is ironic that such joyous songs were the Gershwins' last.
Their final hit, "Love Is Here to Stay," from the *Goldwyn Follies*,
even places love above geological change:

> In time the Rockies may crumble,
> Gibraltar may tumble
> (they're only made of clay)
> but—our love is here to stay

According to Edward Jablonski and Lawrence D. Stewart, it was
Ira who urged George to put in the *and* notes, for they were to be
the song's trick:[9]

> the radio *and*
> the telephone *and*
> the movies that we know

may just be passing fancies—
and in time may go

Such a trick, however, was, like most of Ira Gershwin's, a raggedly mosaic fit that highlighted one of the most ordinary words
in the language by deftly placing it on a strong musical accent.
But before the song was finished, George Gershwin began complaining of excruciating headaches, and on July 9, Ira rushed
him to the hospital for surgery on a brain tumor. Before he lost
consciousness, George whispered a last word to his brother,
"Astaire," a name that epitomized their style of nonchalant elegance, the style of the golden age itself, an age that had already
begun to wane.

Although he continued to write lyrics for over twenty years, Ira
seldom recaptured the slangy sophistication of the popular songs
he had written with his brother. At first he was so devastated by
George's death he did not even contemplate songwriting (Richard Rodgers originally considered asking him to work on *Oklahoma!* but felt Ira was already in "semi-retirement"). When he did
begin working with other composers, such as Kurt Weill and
Jerome Kern, his lyrical style shifted away from vernacular ease
toward the poetic heights. With Kurt Weill, for example, he
wrote songs for a *Lady in the Dark* (1941), as well as for an operetta, *Firebrand of Florence* (1945), where the songs were so integrated into the story that Ira shared billing for the libretto.
   Instead of trying to "write hits" with Weill, he reveled in dazzling rhyming feats, like "Tschaikowsky," a rapid-fire list of the
names of fifty-one Russian composers (which Danny Kaye could
rattle off in record time). There were also witty narratives, like
"The Saga of Jenny" who "in twenty-seven languages . . . couldn't
say no." Occasionally he enlivened his patter with his beloved
contractions and clipped syllables, but the lyrical style was still
Savoyard. *Lady in the Dark* even contained a trial scene with lines
like "for she never promised she wed would" and overt allusions
such as "What do you think this is, Gilbert and Sellivant?"

Gershwin and Weill came closest to having a song detach itself from a show and become an independent hit with "My Ship," from *Lady in the Dark*. The lyric, however, exudes a sentimentality, unrelieved by wit or insouciance, that Ira seldom fell into with his brother. Its uninventive catalog of images—"sails that are made of silk," "a million pearls," "rubies fill each bin," "sapphire sky"—culminates in a cloyingly repeated plea for the ship to bring "My own true love to me." So tangled does Gershwin get in the extended image he weaves that he commits unheard-of poetic inversions, from—"And of jam and spice there's a paradise in the hold" to the stilted—"if there's missing just one thing." When Hollywood made a film out of *Lady in the Dark,* they kept Weill's melody as background music but cut Gershwin's lyric—despite Ira's protests that "My Ship" was so "integral" that the script still carried numerous references to it.

If Gershwin puzzled over Hollywood's fickle tastes then, he must surely have wondered why the Academy nearly gave him an Oscar three years later for "Long Ago and Far Away" (1944). This time his collaborator was Jerome Kern, and, like most lyricists who worked with the old master, Ira Gershwin succumbed to poetic strain. After throwing out one version after another, Gershwin was still stumped. At that point, he recalled, the phone rang and he had to come up with a lyric on the spot for the producer. He took the current version, even though it seemed "just a collection of words adding up to very little," and read it over the phone, secretly relieved "at not having to go down to the studio to face anyone with this lyric."[10] It's hard to believe that the lyricist who wrote " 'S Wonderful" and "How Long Has This Been Going On?" could stoop to such soaring banalities as

> Long the skies were overcast,
> but now the clouds have passed:
> you're here at last!

The song manages only a single instance of Gershwin's usual wordplay—punning on *long* with "that all I longed for long ago was you." While Ira himself loathed it, Kern—and practically everybody else—loved it; "Long Ago and Far Away" didn't win the Oscar, but, ironically, it was the biggest-selling song Ira Gershwin ever wrote.

Ira Gershwin recaptured his slangy brilliance in one last hit song—and another near-miss at an Oscar—"The Man That Got Away," from the 1954 film *A Star Is Born*. This time, however, his collaborator was Harold Arlen, and they were writing for Judy Garland, who became closely identified with the song for ever after. For Arlen, "The Man That Got Away" took him back to his traditional best, like "Stormy Weather" and "Blues in the Night," but for Ira Gershwin "The Man That Got Away" was a new challenge—a full-blown torch song, unusually long (60 measures) and intricately structured (ABABCAD). To meet it, he went back to the vernacular, and came up with one of his cleverest transformations of a catch-phrase (the angler's "one that got away"). Although grammar-minded critics mistakenly kept calling it "The Man Who Got Away," the poetry, typically for Gershwin, was all in the "that," which bitterly reduces the lost man to a mere thing even as it longs for his return.

That bitterness runs throughout the song as Gershwin takes the most hackneyed romantic images and laces them with harsh alliteration to match Arlen's biting musical accents:

> The night is bitter,
> the stars have lost their glitter;
> the winds grow colder,

Then, just as the natural images seem about to continue, Gershwin makes an abrupt shift—not logical but psychological:

> and suddenly you're older—
> and all because of the man that got away

Gershwin uses "and" here with the skill of Hemingway, the simple conjunction linking logically discontinuous—but emotionally connected—ideas.

The rhymes are either starkly simple (bitter/glitter) or imbedded in catch-phrases ("the writing's on the *wall*, the dreams you dreamed have *all* gone astray" or "the man that *won* you has *run* off and *un*done you"). Some are by-products of his characteristic play with contractions and suffixes:

> The road gets rougher,
> it's lonelier and tougher

Here "It's" rhymes roughly with "gets," and the awkward "lonelier" (instead of "more lonely") is forced to rhyme with "rougher" and "tougher." At the climax, he seems to be building up double rhymes—"you burn up," "tomorrow he may turn up," but then he drops rhyme for a prosaic wail whose poetic power comes from two colloquial compounds: "There's just no *let-up* the *livelong* night and day."

Arlen's musical style, while distinctively his own, was nevertheless closer to that of George Gershwin than anyone else Ira worked with. Arlen recalled how easily Ira, who usually found the title the hardest part of lyric writing, came up with the one for "The Man That Got Away." The first time Arlen played the melody for him, Ira leaned over the piano and whispered, "the man that got away," to which the composer murmured, "I like." Perhaps the man that got away, whose absence inspired this heartbroken lyric, was Ira's own long-lost collaborator.

7

The Tinpantithesis of
Poetry: Cole Porter

*Cole Porter is definitive of an era. He IS those years, you
know? He is the style of all those shows, all that period.
He represents it better than anybody else, better even than
Kern or Berlin. Porter's so . . . thirties!*

JOHNNY MERCER

One night in 1925 Cole Porter and some friends were amusing
themselves by improvising parodies of current Tin Pan Alley
"sob-ballads." All of a sudden, one of the friends recalled, Porter's "face grew somber and he said: 'But do you know? I wish I
could write songs like that.' "[1] At the time, such a longing was
understandable; Porter had been writing songs for over a dozen
years without a real hit. Yet even after success came, with the
insouciant "Let's Do It" in 1928, Porter's longing to write such
lachrymose doggerel persisted and lasted to the very end of his
career.

Thus of all the songwriters of the golden age, none presents
more of a paradox than Cole Porter. On the one hand his lyrics,
even more than Hart's or Gershwin's, epitomize an era when
popular song radiated the qualities of *vers de société*, particularly
his "list" songs, which rattle off stupendous catalogs of witty
images and allusions that keeping "topping" each other in verve
and brilliance. On the other hand, Cole Porter wrote some of the
worst lyrics—melodramatic, histrionic, banal—of the age. While

153

Lorenz Hart and Ira Gershwin almost never slipped into sentimentality, Porter frequently wallowed in it.

Charles Schwartz, the most probing of Porter's numerous
biographers, confesses his inability to "reconcile Cole's unmistak-
able maudlin streak," as represented by such "mawkish, heart-
on-the sleeves" songs as "Old Fashioned Garden" and "Hot-
House Rose" with "his even more pronounced unsentimental
one" that sparkles in "such refreshingly cool and sophisticated
gems" as "Let's Do It," "You're the Top," and "Anything Goes."
Citing instances of Porter's love of "tear-jerking insipidities" in
life as well as in song, Schwartz finds such sentimentality "especially surprising in one normally considered the personification
of worldliness, savoir-faire, and even cynicism."[2]

Given his extensive research into Porter's homosexuality, it is
surprising that Schwartz does not speculate about Porter's possibly ambivalent attitudes to the theme of romantic love. If in some
songs, he could look upon such gamboling with humorous detachment or even witty disdain, in others he seems to strain to
write "straight" paeans. Such strain, however, must have been
largely self-imposed, for songs aimed at Tin Pan Alley's popular
market, as we have seen, called for androgynous lyrics—a gen-
derless "I" cooing to an equally indeterminate "you"—that could
be performed by either male or female vocalists, ensuring their
widest commercial dissemination. Still Porter frequently pushed
beyond that formula to write about "True Love" between "honey-
mooners at last alone," turgidly pledging "love forever true."
When he took advantage of the sexual neutrality of the Alley
formula, however, Porter could give an urbane twist to love,
celebrating it as a coolly sensuous—but passing—affair, love as
"just one of those things."

The fact that Porter actually *aspired* to write romantic schmaltz
may also reflect his anomalous position on Tin Pan Alley. Although he is frequently compared to Irving Berlin as one of the
few successful songwriters of the golden age who wrote both
words and music, the comparison ends there. Berlin, the archetypal Alley songwriter—the poor Jewish immigrant boy clawing

his way to the Alley via the Lower East Side—seems perfectly at ease in straightforward Alley fare. Porter, however—son of wealthy midwesterners, educated at Yale and Harvard, darling of European and American cafe society—is more at home in the flippantly sophisticated lyrics he performed at elegant parties long before they brought him popular acclaim. It was Elsa Maxwell, the center of that international set, who consoled a frustrated Porter in the early 1920s by telling him "your standards are too high. The wit and poetry of your lyrics are far beyond the people. But one day," she prophesied, "you will haul the public up to your own level."[3]

Yet it was success not just on his own terms that Porter wanted but on the Alley's terms as well, an Alley where he was long dismissed as a wealthy dilettante. He clearly longed to write *popular* songs, and that longing is reflected in his famous remark to Richard Rodgers, in 1926, that he had at last found the secret for writing hits. "As I listened breathlessly for the magic formula," Rodgers recalled, Porter whispered that it was "simplicity itself: I'll write Jewish tunes." As Rodgers interpreted that remark in the light of the successful songs that soon followed, it signaled a shift toward minor-key, heavily chromatic music that sounded "unmistakably Mediterranean":

> It is surely one of the ironies of the musical theatre that despite the abundance of Jewish composers, the one who has written the most enduring "Jewish" music should be an Episcopalian millionaire who was born on a farm in Peru, Indiana.[4]

The lyrical equivalent for writing "Jewish" music seems to have been to strain for maudlin ardency, and his most strident lyrics, like the "beat, beat, beat of the tom-tom" in "Night and Day" are often fitted to his most brooding chromatic melodies.

Those melodies, moreover, frequently overrun the bounds of the standard Alley thirty-two-bar AABA formula (the chorus of "Night and Day," for example, runs to forty-eight measures), and the more convoluted musical pattern may have further induced Porter to dampen his clever rhymes and dazzling imagery so they did not overshadow the music. Thus he almost never uses his patented "list" technique in such songs, reserving those witty catalogs for melodies that follow the more simple and standard musical pattern, such as "Anything Goes" and "You're the Top." Lyrically speaking, Porter was at his urbane best when toting up a clever inventory; when he abandoned the list, he frequently slipped into sentimentality.

Whatever the cause of Porter's paradoxical shifts between sophistication and schmaltz, they are apparent from his very first lyrics, written for student musicals at Yale. On the one hand there are many excellent comic songs, frankly derivative of Gilbert and Sullivan patter, such as "When I Used to Lead the Ballet" (1912), where an aging "Premier Danseur" recalls his gorgeous costumes,

> In the get-up I had on
> I looked like the "ad" on
> a bottle of sparkling White Rock!

and the chorus chips in, Savoyard-fashion, with

> He was paid to say that,
> He was paid to say that,
> He was paid to say that about White Rock

In another Yale musical, from 1913, we hear his characteristic blend of sophisticated diction and colloquial ease, as well as his equally characteristic juxtaposition of images from high and low culture in a song for a chorus of impoverished college boys:

> Death hangs o'er us
> like the threatening sword of Damocles,
> we're so poor
> we can't afford a box of Rameses

Interspersed among these, however, one also finds love songs decked out in the tritest Alley formulas:

> Goodbye, my true love,
> fare thee well, mine own.

> I'll dream of you, love—
> none but you alone.

Porter's stylistic schizophrenia is equally apparent in the songs from his first Broadway show, in 1916, *See America First*. On the one hand, it featured the kind of allusive, incongruous "list" song that looks forward to "You're the Top." "I've a Shooting Box in Scotland," for example, rattles off a catalog of elegant objects— "prints by Hiroshigi, Edelweiss from off the Rigi, Jacobean soup tureens, early types of limousines"—interspersed with prosaic American allusions to Ty Cobb and "Jeff and Mutt." Yet when he turned to declarations of love, his language became ardent,

> Ah, worse than death the plight
> of heart for heart repining,

"poetic,"

> Naught shall sever us now,
> That is my vow most true,

and convoluted,

> From far thou dost bring
> on magical wing
> a memory ever dearest.

Partly because of such songs, which reviewers found "unimpressive," but also because of poor production and an inexperienced cast, *See America First* closed after only fifteen performances. That the failure of the show inspired him to join the French Foreign Legion was one of Porter's many flamboyant fictions. For the next few years, however, he kept his distance from Tin Pan Alley. After a marriage of convenience to heiress Linda Lee, he lived the lavish expatriate life in Europe. In Paris, Venice, and London he found an enthusiastic private audience for his witty songs in an international set that included Noel Coward, Gerald and Sara Murphy, and Elsa Maxwell. Maxwell recalled parties where Porter performed his witty songs, some of

the same ones that had flopped in *See America First,* to an "enrap-
tured" audience, "straining to catch the droll nuances of his
lyrics."

Porter's second chance at Broadway came in 1919. On a
voyage back to America he met producer Raymond Hitchcock,
played some of his songs, and was commissioned to do the score
for the third edition of Hitchcock's annual *Hitchy-Koo* revues. In
New York, Hitchcock also introduced Porter to Max Dreyfus,
head of T. B. Harms, one of Tin Pan Alley's biggest publishing
houses. Although the reviewers praised *Hitchy-Koo of 1919*—
singling out Porter's "particularly clever job of the lyrics"—the
show closed early and none of those clever songs became popular.

The one song that did sell substantial copies of sheet music,
"Old-Fashioned Garden," must only have comfirmed Porter's belief that popularity came from clichéd rhymes,

> One summer day I chanced to stray

mildewed imagery,

> to a garden of flow'rs blooming wild

poetic diction,

> It took me once more to the dear days of yore

and unabashed nostalgia:

> and a spot that I loved as a child.

Only in such a lyrical setting, "in the land of long ago," far from
the homosexual entourage that surrounded the Porters in Europe, could love be declared to "an old-fashioned missus" by
"an old-fashioned beau." "Old-Fashioned Garden," was, moreover, a classic written-to-order Alley song, the kind Porter
longed to write: Hitchcock had bought a set of second-hand
flower costumes from Ziegfeld and ordered Porter to come up
with a flower song so he could use them in a production number. As Charles Schwartz notes, however, Porter's botanical
knowledge was so slight that he had "phlox, hollyhocks, violets,

eglantines, columbines, and marigolds all blooming at the same time"; nonetheless, "the American public, responding to the gushy sentiments in it, bought thousands of copies of sheet music."[5]

Occasionally Porter managed to get one of his witty songs, such as "Olga, Come Back to the Volga," interpolated into an English or American revue, but the songs themselves never became popular. In 1922 he wrote another *Hitchy-Koo* replete with a sentimental "Old-Fashioned Waltz," a clone of "Old-Fashioned Garden" right down to "the days of yore," but this one never even made it to New York, closing in Philadelphia after less than two weeks. Once more Porter withdrew to Europe, ensconced himself in a Venetian palace, and tried his hand at serious music by collaborating with Gerald Murphy on a jazz-ballet, *Within the Quota.*

Once more, however, a chance encounter brought him back to Broadway, when director John Murray Anderson, upon hearing Porter play his songs at a party in Paris, invited him to provide songs for the 1924 *Greenwich Village Follies.* Still Porter continued to divide his style between urbanely comic songs and banal love ditties. On the one hand, there was the racy narrative of "Two Little Babes in the Wood," who learn that "the fountain of youth is a mixture of gin and vermouth." "I'm in Love Again," on the other hand, is tepid Alley fare, replete with "Spring is comin' " and "heart strings strummin'." While *Greenwich Village Follies* was a success, Porter's songs were not; in the course of its New York run, the revue dropped his songs one by one, and by the time it went on the road, all of Porter's songs had been cut.

Thus one can easily understand Porter's saying, in 1925, that he would give anything to be able to write a Tin Pan Alley "sob-ballad." The failure of his songs for the *Follies* deepened his sense that his songwriting career was doomed. Friends tried to console him by assuring him his songs were too stylish for popularity, a judgment echoed by the music critic Deems Taylor, who, upon hearing Porter play some of his flops at a New York party, pronounced the "lyrics much too sophisticated to have any popular success."[6]

Yet even as they spoke, popular taste was changing as a result of the success of such sophisticated theater songs as those of Rodgers and Hart and the Gershwins. While such songwriters were paving the way for public acceptance of Porter's urbane songs, he diligently kept working, nurturing his own distinctive blend of vernacular ease and witty elegance that would epitomize the golden age. Finally, in 1928, he had his first successful show with *Paris,* and his first genuine hit song, "Let's Do It." Designed as a showcase for Irene Bordoni, the "French Doll" of Broadway, *Paris* was the perfect vehicle for Porter as well, since Bordoni's elegantly naughty delivery was tailor-made for the kind of lyrics that had long titillated his friends.

What Bordoni brought to a lyric, however, was not only insouciance but sensuousness, so much so that when she sang Buddy DeSylva and George Gershwin's "Do It Again" (1922), those three little words radiated such erotic force that the song was banned from the radio. Similarly, "Let's Do It" introduced a new passionate urgency into Porter's witty catalog. That erotic energy can be seen by comparing "Let's Do It" with a similar song, "Let's Misbehave" (1927), originally planned for *Paris* but replaced by "Let's Do It." Set to a sprightly melody, "Let's Misbehave" proffers a list of traditional pleas for "Un peu d'amour," such as the *carpe diem* exhortation that we should make much of time "while we're still active." One of those standard pleas invokes natural creatures who gambol in the spring and argues that we should "naturally" follow suit:

> They say that bears
> have love affairs,
> and even camels;
> we're merely mammals,
> let's misbehave.

Such copulating creatures are evoked gingerly, however, and our dalliance is urged in the most discrete terms, the arch language disclosing sly puns:

> They say that spring
> means just one thing
> to little love birds,
> we're not above birds,
> let's misbehave.

Except for these pleas to emulate our fellow-creatures, however, the rest of the lyric presents fairly routine urgings that depart from the colloquial flow in such cumbersome convolutions as "You know my heart is true, and you say you for me care."

Where "Let's Misbehave" pleads coyly, "Let's Do It" seethes with a cool urgency signaled by the shift in the title from the quaint "misbehave" to the simple forthrightness of "do it." The utterly innocent vernacular phrase is transformed into the bluntest of propositions simply by placing it in a love song, though the chromatic drop to a G flat on "do" is a musical off-color wink that, as Alec Wilder notes, "gives immediate character to the melody."[7] Although it is enlivened by clever leaps and descents, that melody, couched in the standard AABA formula, consists of brief patterns of repeated notes, and Porter shrewdly uses that musical insistence to underscore the urgency of his plea.

What marks "Let's Do It" as Porter's first major lyric, then, is that it uses the technique of cataloging not just as a framework for witty images but as a device for heightening passionate intensity. Just as Lorenz Hart employed lacerating rhymes to wryly register the pain of romance and Ira Gershwin used linguistic playfulness to capture the euphoria of first love, Porter used the "list" to underscore the erotic intensity of his romantic pleas and celebrations. In "Let's Do It" his listing of various creatures and their modes of copulation mirrors the very erotic universe it describes—image propagating image with an imaginative fecundity that rivals nature's own fertility.

Taking his cue from the gamboling bears and camels of "Let's Misbehave," Porter devotes "Let's Do It" to a sexual bestiary, each refrain devoted to a different biological order. The first refrain inventories humans who "do it," but, with great tolerance and good humor, invokes largely unromantic types—not the French (whom we know do it), but those we never think of in erotic terms, such as the Dutch, the Finns, the Lapps, the Lithuanians, and, punning on his title, the Letts. In listing instance after instance of universal fornication, Porter uses such genteel catch-phrases as "not to mention" and "I might add," giving the catalog the air of refined gossip.

It's the very gentility evoked by such language that inhibits "doing it," and Porter's list concentrates on examples of how the most prudish and elegant creatures nevertheless give in to the universal urge: "sedate barn-yard fowls" and "high-browed old owls," "refined lady bugs" and "overeducated fleas," "English soles" in "shallow shoals" and discreet goldfish "in the privacy of bowls." In Boston, that citadel of Puritan propriety, "even beans do it." Other reluctant creatures offer still greater testimony to the universal and irresistible power of *eros:*

> Cold Cape Cod clams,
> 'gainst their wish,
> do it,

and even "lazy jelly-fish" manage to summon the energy to fornicate, as do heroic electric eels, "thought," Porter politely puns, "it shocks 'em, I know."

However, it is our closest kin, the mammals, who must overcome the greatest obstacles of all: "heavy hippopotami," embarrassed giraffes, imperfectly ballasted kangaroos, and "old sloths who hang down from twigs" all manage to "do it, though the effort is great." In fact, it is only through the staunchest of Victorian virtues—earnestness, endurance, and persistence—that such creatures reproduce themselves. Thus the strategy of the list appeals to the very values that would oppose "doing it," and the relentless energy that spawns the extensive catalog is itself testimony to the creative stamina that animates the erotic universe.

Part of that generative force is Porter's clever word-play, propagating puns out of the simplest terms: "Larks, k-razy for a lark, do it," "So *does* ev'ry katy*did do* it," and

> Moths in your rug, do it,
> what's the use of moth balls?

Such sly testicular punning made this lyric one that had to be listened to with the same attentiveness readers gave to the nuances of light verse. The fact that such a witty song could not only succeed on the Broadway stage but achieve independent popularity in Tin Pan Alley gave further credence to the lyrical "renaissance" inaugurated by Rodgers and Hart's "Manhattan." After years of disappointment, Porter was hailed as one of a new breed of "sophisticated" theater writers who could write "hits." Reviewers praised his "delicate balance between sense, rhyme, and tune" that "makes other lyrists sound as though they'd written their words for a steam whistle."

The following year, 1929, Porter had another successful musical, *Fifty Million Frenchmen,* where he showed that he could write not only clever catalogs but a "straightforward love song" without lapsing into his turgidly sentimental style. Alec Wilder praises "You Do Something to Me" as "musically and lyrically without guile or cynicism," much like a Kern song except for "an unexpected rhythmic stunt" that is "unmistakably Porter."[8] That rhythmic turn is fitted to an equally characteristic lyrical twist that registers romantic affect yet treats it with urbane, almost clinical, detachment:

> do do
> that voodoo
> that you do
> so well.

Yet that detachment has an urgency, which Porter underscores not only with rhyme, each *oo* rhyme falling on a different note as the melody descends, but by another of Porter's lyrical devices that match his musical chromatics—repeating the same word but shifting its meaning as he does with *do* in "do do" and *that*, using it first as an adjective ("*that* voodoo"), then a relative pronoun ("*that* you do so well").

In other songs, however, he could not repress his longings to write brooding, melodramatic ballads, particularly when he had to set more heavily chromatic melodies. "What Is This Thing Called Love?" (1929), for example, shifts dramatically between major and minor keys, and Porter fits such tonal ambiguity to a lyric of tormented bewilderment. Yet except for taking a clichéd metaphor,

> you took my heart

and giving it a brutal, Hart-like twist,

> and threw it away

the lyric never upstages the music. Instead, Porter tries to find lyrical equivalents for the musical chromatics—repeating the same sound with different meaning in "*one won*derful day," for example, or twisting the vowel of "saw" through "called" and "Lawd." Some of the lyrical histrionics may reflect the fact that, in the revue *Wake Up and Dream*, "What Is This Thing Called Love" was sung while dancer Tillie Losch, backed by tempestuous tom-toms, gyrated before an African idol.

Those tom-toms must have reverberated in Porter's imagination when he wrote "Night and Day" (1932), his most famous song and one that was so popular it made a hit out of an otherwise weak production, *Gay Divorce*, which became known as the "Night and Day" show. The song starts off "Like the beat, beat, beat of the tom-tom when the jungle shadows fall," and while heavily chromatic music oscillates for forty-eight measures, the lyric shifts melodramatically between imagistic "chromatics": moon and sun, near and far, and, of course "night and day." Once again, Porter suppressed wit and cleverness, bringing out the musical intensity with tiny, faceted repetitions of sound—one/*under*/*sun*/*hun*gry, m*oon*/r*oom*/b*oom*/thr*ough*/y*ou*.

Such sound effects brought high praise from Irving Berlin, that master of faceted sound, who thought "Night and Day" Porter's "high spot." The lyric sounds like a Berlin ballad, right down to "the silence of my lonely room," though where Berlin might have located a "hungry yearning burning inside of me" in the heart, Porter is more his own sensuous self when he puts it "under the hide of me" (a forecast of "I've Got You Under My Skin"). Porter, too, is closer to Hart than Berlin when he emphasizes the physical torment of love, though it's hard to imagine Hart portraying that torture in such melodramatic similes as

> the tick tick tock
> of the stately clock
> as it stands against the wall

or

> the drip drip drip
> of the raindrops
> when the summer show'r is through.

One can hardly blame Ring Lardner for his lisping parody: "Night and day under the fleece of me, there's an, oh, such a flaming furneth burneth the grease of me."

Porter's high spots came when he stuck to simple musical structures and complex lyrical catalogs. In the mid-1930s, he wrote a number of dazzling list songs for a series of Broadway shows—*Anything Goes, Jubilee, Red, Hot, and Blue*—that vie with those of Rodgers and Hart during the same years as the epitome of the golden age. Not only were these songs successful on stage, they, like Rodgers and Hart's, detached themselves to become independent hits. (*Anything Goes* alone produced five major hit songs.) In these songs, as in "Let's Do It," it is not just the clever items in the catalog but the act of listing itself, the energy that spawns and sustains those items, that is as much the subject as the beloved it celebrates, cajoles, or laments.

"You're the Top" (1934), for example, bubbles over in a rapid-fire list of images and allusions that threatens to go on endlessly (indeed Porter kept adding refrains over the years). If such catalogs have an analogy in modern poetry, it would be to e. e. cummings' poetic collages, which satirically unite "Abraham Lincoln and Lydia E. Pinkham" with "Cluett/Shirt Boston Garter" and "Spearmint Girl with the Wrigley Eyes." "You're the Top," similarly, juxtaposes images of European high art—the Louvre Museum, the Mona Lisa, a Shakespeare sonnet, a "symphony by Strauss" (who never wrote one)—with some uniquely American artifacts—a Bendel bonnet, a "Berlin ballad," Cellophane, Mickey Mouse. Yet as in collage, where apparently discordant elements turn out to have underlying resonances, some of Porter's items straddle both worlds: "You're Whistler's Mama," for example, slangily invokes the common name for a painting by an American expatriate that hangs in the Louvre (and ranks second only to the Mona Lisa on the tourist's "must see" list). Similarly,

> you're a rose,
> you're Inferno's Dante,
> you're the nose
> on the great Durante

at first seems to juxtapose European high-brow against American low until, as noted earlier, we see that both Dante and Durante are Italian comedians, an identity underscored by the triple rhyme: *rose, nose,* Infer*no's*. So, too, "Mrs. Astor," from the aristocracy of what was once New Amsterdam, is not unlike an "old Dutch master," which, in turn, denotes not only a painting but a cigar, a product as commercial as European camembert. In the potpourri of "You're the Top," the very juxtaposition of images blurs their distinctiveness, so that "Mahatma Gandhi" sounds as much like a product as "Napoleon Brandy." Similarly, a list of things "diveen,"

> You're a Botticelli,
> you're Keats,
> you're Shelley,
> you're Ovaltine

makes all four sound equally like paintings, poets, or products. Such a brilliant list testifies to the fecund imagination of a lover who, despite his protests of being the "bottom," perpetually out-tops himself in an orgy of inventiveness.

In another catalog duet, "It's De-Lovely" (1936), the erotic energy of the list is even more explicitly connected with the natural fecundity it celebrates. The verse expresses a "sudden urge to sing" that is part of the universal fecundity of nature in the spring, and the same creative urge manifests itself in the language of the song as it moves increasingly toward fertile abandon. In the verse language is being pushed to earthier simplicity, as one singer daintily invokes "that certain je ne sais quoi" and the other retorts, " 'stead of falling into Berlitz French, just warble to me, please" in "plain Brooklynese."

The chorus carries this slanginess further and further, as a catch-phrase like "You can tell at a glance" is followed by "what a swell night this for romance" and "I understand the reason why you're sentimental" is offset by " 'cause so am I." The list itself consists, not of images or allusions, but of permutated prefixes that eventually transform the "de-" of such perfectly proper ejaculations as

> It's delightful, it's delicious,
> it's delectable, it's delirious

into the Brooklynese grunts of "it's delimit," and "it's deluxe" (which Porter specified should be pronounced "de-lukes").

Porter may have been responding to Ira Gershwin's slang play as he carried out his linguistic list for five more refrains, through "divine, diveen, de-wunderbar, de victory, de vallop, de vinner, de voiks," then "de-reamy," "de-rowsy," "de-reverie," "de-rhapsody," "de-regal," "de-royal," "de-Ritz," on to "appalling," "appealing," "a pollywog," "a paragon," "a Popeye," "a panic," "a pip," and finally "a failure," "a fold-up," "a fadeaway," "a fare-thee-well," "affliction," "a flaccus," and finally suspending the prefix itself as the song ends on an alphabetically originary "a-." In a lyric that celebrates the inexhaustible fertility of "Mother

Nature" in the spring ("murmuring low, 'Let yourself go!' "),
Porter's cataloging matches nature's procreative energy with
equally earthy—and seemingly endless—verbal fecundity.

Even more linguistically abandoned is the catalog of "Anything Goes," a song which, like Berlin's "Puttin' on the Ritz," uses
the music's rhythmic pulses to so rag lyrical accents that at times
the words are incomprehensible. A phrase such as "When folks
who still can ride in jitneys find out Vanderbilts and Whitneys
lack baby clothes, anything goes," comes out rhythmically as

> When *folks* who *still*
> can *ride* in *jit-*
> neys find *out* Van-
> der*bilts* and *Whit-*
> neys lack *ba-*
> by *clo'es*

Even more incomprehensible is "When *every* *night* the *set* that's
smart *is* intr*ud*ing in *nud*ist *par*ties in *stu*dios," a line that some
critics have found "*too* clever, too 'lyric-y.' "[9]

Such radical ragging even distorts the images in the catalog:

> that *gent* to-
> day you gave a *cent* to-
> day once had several châteaux.

Here the rhythm detaches "to" from "today" to create the mo-
mentary ambiguity of "gave a cent to," thus adding still more
semantic confusion to a lyric that exemplifies the principle of
"anything goes." That title itself flip-flops in meaning, from the
ballyhoo years of the 1920s when "anything goes" was a slogan of
moral abandon, to the 1930s when the phrase became a bleak
reminder of the stock-market crash.

Songs like "You're the Top," "It's De-Lovely," and "Anything
Goes" use a copious catalog of images, allusions, and verbal transformations to underscore passionate urgency, an urgency, however, that is always urbanely muted by the witty items in the list. In

other songs, Porter uses a list of only a few images, developing the tension between passion and sophisticated detachment through character, creating a figure who is as much his own distinctive transformation of the standard Alley lover as Hart's wry masochist or Gershwin's euphoric innocent. Porter's characters strike a pose of bemused sophistication, trying to understate both euphoria and heartache, yet that veneer of cool sophistication cracks, if ever so slightly, to reveal a brief flash of real feeling.

Such a character, for example, may face the end of an affair by dismissing love as "Just One of Those Things" (1935), a wryly flippant slang formula that makes parting not sweet sorrow but casually commonplace—"just one of those" (and one of those favorite Porter words) "things." While he trivializes the notion of eternal love, however, the catalog of images undercuts his insouciant pose; how, for example, can a

> trip to the moon on gossamer wings

be

> just one of those things?

Even the images that seem ordinary are transformed by deft phrasing: "one of those bells that, now and then, rings." Here the "now and then" undercuts the singer's seemingly flippant disregard, and when the phrase reappears at the close of the song— "Here's hoping we meet now and then"—it suggests, in another brief flash of feeling, that the singer might genuinely want to express sorrow but cannot doff his urbane mask.

"I Get a Kick Out of You" (1934) presents the same Porter lover "fighting vainly the old ennui" and complaining "practically ev'rything leaves me totally cold":

> I get no kick from champagne;
> mere alcohol doesn't thrill me at all

As so often in Porter's best songs, slang—"I get no kick"—is offset by archly elegant diction: "Mere alcohol." Yet such laconic hyperbole serves as an offhanded and understated tribute to the power of the beloved: "so tell me why should it be true that I get a kick out of you?"

Porter further disrupts the singer's mask of controlled ennui by his skillful use of triplets that throw the lyrical phrases off balance, breaking it up into such shards as "-hol-does-n't" and "thrill-me-at." In the second A section, a more potent image, cocaine, is introduced and, with it, a more tipsy disruption of the claimed imperviousness:

> if-I-took
> e-ven-one
> sniff-that-would
> bore-me-ter-
> rif-ic'-ly

The imbalance here is heightened by the triple rhyme—if/sniff/ 'rif—which Porter places on the first note of alternating triplets. So intricately off-balance are Porter's words and music that some singers miss the ironic point and even out the triplets into a waltz rhythm. Astute singers like Ethel Merman, however, emphasized Porter's irony by inserting a slight but lingering pause in the midst of "rif-ic'-ly" to suggest the heady effect of one whiff of the supposedly ineffectual cocaine.

In the release Porter moves away from his list of intoxicating imagery and has his composed singer not only declare his feeling but indicate that it is thoroughly unrequited: "I get a kick though, it's clear to me, you obviously don't adore me." Such an urbanely casual admission of the hopelessness of his adoration provides a different gloss to the "kick" he receives—not an exhilarating high but a cool rebuff that, nonetheless, leaves him masochistically longing for more. Thus the character who emerges from the lyric, far from being impervious to sensation, is hopelessly addicted to his lover's "kick," an addiction all the more poignant given his façade of nonchalant sophistication.

In the final A section, Porter alters his imagery: rather than trying to top his list with a third, more potent drug than alcohol or cocaine, he takes the "high" they give in literal terms and underscores the ascent with a series of rhymes on the progressively higher notes of the melody: "flying too high with some guy in the sky is my idea of nothing to do." Despite the imagistic shift, Porter laces his three "kickers" together with rhyme: champagne, cocaine, plane.

Amidst all of these brilliant catalog songs of the 1930s, Porter was still grinding out lugubrious sob-ballads. The brooding hit from *Anything Goes* was "All Through the Night," a thoroughly chromatic melody running to sixty-four measures, which, for Alec Wilder, barely misses becoming "an arty piece of melodrama."[10] The lyric, however, succumbs. As in "Night and Day," Porter lets the music determine the words; its relentless, shifting melody spawning a lyric devoted to the "monotone of the evening's drone," using echoing rhymes rather than witty images:

> All through the night, under bright stars above
> you and your love will bring ecstasy.
> When dawn's overtaken us, we'll sadly say goodbye,
> till dreams reawaken us and the moon is high

Right down to its obligatory high moon and bright stars, the lyric retools the tone, the imagery, the setting of "Night and Day," as if such music automatically engendered the same constellation of lyrical effects.

That heavenly cluster turned up again in "Begin the Beguine" (1935), at a whopping one-hundred and eight measures one of the longest popular songs ever written. Those long melodic lines pushed Porter to new celestial heights. Once, on "a night of tropical splendor" lovers dance "under the stars" and know (poetically inverted) "moments divine" and "rapture serene," till the inevitable "clouds came along to disperse the joys we had tasted." The mixed metaphor somehow leads to the prolonged reflection.

> And now when I hear people curse the chance that was wasted,
> I know but too well what they mean.

But Porter—and his astronomically long song—are not dead yet: despite his plea "let it sleep like the dead desire I only remember," in one last gasp he begs for the beguine (One more time!) until "the stars that were there before return above you." While "All Through the Night" barely escaped artiness for Alec Wilder, he considers "Begin the Beguine" musically so turgid that "along about the sixtieth measure" he finds himself pleading "End the Beguine."[11]

Porter piled the melodrama and sentimentality even higher when he wrote songs for the movies. In one of his first Hollywood songs, "Good-bye, Little Dream, Good-bye" (1936), he displayed his usual minor-key wares for MGM's executive producer, Sam Katz, along with the purple poetry that went with it:

> For the stars have fled from the heavens,
> the moon's deserted the hill,
> and the sultry breeze that sang in the trees
> is suddenly strangely still

Katz's reaction must have confirmed Porter's belief that he had indeed discovered the key to success: "You know, Cole," he said, "that song is beautiful, it's—why, it's Jewish."[12]

He had an even greater triumph with "In the Still of the Night" (1936), a song even Nelson Eddy (Nelson Eddy!) found overblown and at first refused to sing in the film *Rosalie.* It is inordinately long (seventy-two measures) and has a complex structure that includes an "agonizing" sixteen-measure stretch marked "Appassionato," but which Wilder describes more pithily as "bathos, pure and simple."[13] Yet when Porter played "In the Still of the Night" for Louis Mayer, the most rugged studio mogul of them all, Mayer wept. "Imagine making L. B. Mayer cry," he told a friend, "What could possibly top that?"[14] For the despondent composer who in 1925 moaned that he would give anything to write the "sob-ballads" he despised, Mayer's lachrymose response was the gusher that finally came in. One can imagine Porter milking those tears as he lingered over such tremulous rhymes as

like the moon growing dim
on the rim
of the hill
in the chill,
still
of the night

On Mayer's orders, Nelson Eddy sang the song.

Yet while Mayer sobbed over "In the Still of the Night," he grumbled that the title song for *Rosalie* was "too high-brow." "Forget you are writing for Nelson Eddy and simply give us a good, popular song." Porter resented being forced to rewrite, but dutifully "took 'Rosalie No. 6' home and in hate wrote 'Rosalie No. 7.'" Invoking the usual "night, when stars danced above" Porter implored, like any Alley wordsmith,

Won't you make my life thrilling,
and tell me you're willing
to be mine, Rosalie, mine

While some have suggested that he "deliberately set out to write the worst song he could" with "insipid lyrics and an uninspired melody,"[15] "Rosalie" is squarely in his usual sentimental vein. When Porter confided to Irving Berlin how much he hated the song, Berlin quipped, "Listen, kid, take my advice, never hate a song that has sold half a million copies."

Occasionally in his Hollywood songs, Porter could emulate Berlin's own clean simplicty, striking a middle ground between his own extremes of witty urbanity and strident sentimentality. "Easy to Love" (1936), for example, registers the ease of its own title, relying, as Berlin so often did, on the most casual contractions— "You'd be"—and subtle repetitions of vowel and consonant sounds—"easy" echoing the simple "be so" that precedes it. Similarly, in the closing lines,

That you can't see
your future with me
'cause you'd be, oh,
so easy to love

"future" recapitulates the sounds of both "you" and "your."

Another Hollywood song, "I Concentrate on You" (from *The Broadway Melody of 1940*) takes the *on* and *en* diphthongs in "Concentrate *on* you" and has them echo through the chorus— "str*ong*," "s*ong*," and wr*ong*," and the numerous "*on*'s"; then "t*en*der," "surr*en*der," "wh*en*," and "m*en*." Porter also manipulates his vowels as skillfully as Berlin, pairing off *i* and *a* in a sequence of balanced phrases—"sk*i*es look gr*a*y to me," "cr*i*es 'n*a*y, n*a*y,' to me," and "w*i*se men s*a*y to me." Similarly, he weaves *oo* through "trouble begins t*o* br*ew*," "people declare 'You're thr*ou*gh,'" "never comes tr*ue*," and, in the final line, "T*o* pr*o*ve that even wise men can be wrong I concentrate on y*ou*," a line, followed by three repetitions of the title phrase in a closing concentration of vowels that mirrors the singer's intensely focussed ardor.

Probably his most artfully simple film song was the verseless "I've Got You Under My Skin" (1936), its catch-phrase title lifted from its normal context of irritated exasperation ("you're starting to get under my skin") and given a nonchalantly sensuous frame. Then Porter adds a paradoxical touch—what the singer "has" under her skin is not her beloved but a hopeless addiction to a lover she knows she *can't* "have." Yet that very awareness only makes her plight more poignant, as a slangy voice "comes in the night" and "repeats and repeats" in her ear:

> "Don't you know, little fool, you never can win,
> Use your mentality—wake up to reality!"

Yet, like a hopeless addict, she confesses her plight in a reversal of the catch-phrase warning "stop before you begin":

> But each time I do,
> just the thought of you
> makes me stop,
> before I begin,
> 'cause I've got you under my skin.

Paradoxically, she can't "stop," can only "stop" trying to "stop."

Even though he could write such deceptively simple songs for films, the primary showcase for Porter's urbane style remained the theater. But Broadway itself was gradually beginning to change, and that change coincided with a momentous change in Porter's own life. In 1937, soon after he returned east to work on a new musical, he underwent what Brendan Gill has called "the central episode" of his life—"not the most important one but the one that everything else stands in relation to."[16] Out horseback riding one morning, Porter was thrown from his horse, then the horse rolled over him, crushing both his legs. The accident left him crippled and in excruciating pain for the rest of his life, finally necessitating amputation a few years before his death in 1964. Porter told a story about the accident which, whether true or not, epitomizes the debonair style of the age he represented: while he lay on the ground waiting for help, he composed the casually elegant "At Long Last Love."

One of his last great list songs, "At Long Last Love" (1938), takes Porter's characteristically world-weary lover and has him skeptically question his newly-smitten state with a catalog that belies the vulnerability—even innocence—beneath his cosmopolitan veneer:

> Is it an earthquake or simply a shock?
> Is it the good turtle soup or merely the mock?
> Is it a cocktail—this feeling of joy,
> or is what I feel the real McCoy?

All the Porter touches are here—the witty images that range from the European elegance of "Bach" to the prosaically American "Chevrolet," the blend of elevated diction and brash slang in conversational phrasing, and the cleverly skewed rhymes of sh*ock*, m*ock*, co*ck*tail, and Mc*Coy*. As in the best of his list songs, the act of cataloging itself registers passionate hopefulness even as it strives for cool detachment.

While not a swan song, "At Long Last Love" was one of the last of Porter's urbane list songs to enjoy popular success—a sign that, on the eve of World War II, an age of cosmopolitan sparkle was beginning to wane. Porter's accident, as we have seen, coincided with George Gershwin's death and Lorenz Hart's worsening alcoholism; it did not, however, keep him from writing shows. Nonetheless, while the shows themselves were often successful, his ability to write hit songs began to decline. "At Long Last Love" was the only song from *You Never Know* to become popular, and the show itself closed after a disappointing run of only fourteen performances. Much more successful was *Leave It to Me* (1938), yet it too generated only one hit, "My Heart Belongs to Daddy," added as an afterthought to cover a scene change but transformed into a show-stopper by newcomer Mary Martin, who delivered its risqué lyric in an innocent, child-like voice while she did a mock-striptease in furs.

Between 1939 and 1944 Porter had five hit shows, but none of his witty list songs transcended the theater to become popular—not Ethel Merman's sexily antisexual duet with Bert Lahr in *Du Barry Was a Lady*, "But in the Morning? No!" (1939) nor their pledges to one another in "Friendship," such as "If you ever lose your teeth and you're out to dine—borrow mine!" *Panama Hattie* (1940) prompted John O'Hara to lament, "Who'd have thought we'd live to see the day when Cole Porter—*Cole Porter!*—would write a score in which the two outstanding songs are called 'My Mother Would Love You' and 'Let's Be Buddies'? And written straight, too; no kidding." Although O'Hara tried to explain away the schmaltz with an "Ah well, he had a bad riding accident a year or two ago,"[17] both songs simply offer more testimony to Porter's ongoing stylistic schizophrenia.

That schizophrenia is apparent in *Let's Face It* (1941) as well. Catalogs like "Let's Not Talk About Love" delighted theater audiences as Danny Kaye gave rapid-fire delivery to such items as

> Let's ride the New Deal, like Senator Glass,
> let's telephone to Ickes and order more gas

But, as Porter himself admitted, such witty lists now largely appealed only to Broadway "first-nighters." The closest he got to a popular hit was the ditty "Ev'rything I Love," with its clichéd pledges like "you are to me ev'rything, my life-to-be" and "Each time our lips touch again I yearn for you, oh, so much again."

In *Something for the Boys* (1943), Porter's attempt to emulate Irving Berlin's successful soldier show, *This Is the Army*, he didn't even come close to a hit. Reviewers now began to complain that "Cole Porter's last few shows," as measured by "the real test of a musical"—its ability to produce popular songs—"have been most disappointing and this one perhaps most of all. . . . In this one he also fails as a lyricist." The one hit song from *Mexican Hayride* (1944) gave even stronger indications of such failure; Porter's friend Monty Woolley had wagered he couldn't write a song in which the phrase "I Love You" was repeated over and over. Just as Lorenz Hart, on a similar dare, had used every Alley cliché he could think of for "Blue Moon," Porter first had the title phrase hummed by the "April breeze," then echoed by the inevitable hills, and finally reiterated by "the golden dawn" as "once more she sees"—what else?—"daffodils." "I Love You" quickly went to the top of the Hit Parade, but even its popularity was eclipsed when Roy Rogers sang "Don't Fence Me In" in the film *Hollywood Canteen*, a song so successful it spawned its own film in 1945. Because "Don't Fence Me In" seems so unlike Porter, some reviewers assumed he had written it as a parody of Tin Pan Alley cowboy songs. But the song went back to a 1934 film musical that was never produced, *Adios, Argentina;* the lyric was based on a poem by a Montana versifier named Robert Fletcher. Porter had acquired the rights to the poem and reworked them, but shelved the song when *Adios, Argentina* was cancelled. In the 1940s, however, after the regional nostalgia of *Oklahoma!*, its time had come, and with even Cole Porter doffing his top hat for a ten-gallon sombrero, the age of urban sophistication was clearly over.

While the age he epitomized was over, Porter's own songwriting career was not. In fact, his greatest Broadway success was still to come, in 1948, when he showed that he, like Irving Berlin, could write a musical in the new "integrated" style. Until then Porter had been what Gerald Mast described as the "one master who continued to march against the movement of musical-theater history—writing shows for their songs rather than songs for a show."[18] With *Kiss Me, Kate*, however, this songwriter, who long had had blithe disregard for the book of a musical (often dropping off his songs at the theater before rehearsals and then disappearing), found, for the first time, what Martin Gottfried calls a *"musical milieu"* that provided a "unifying style" for his songs and kept them, for once, from wandering "off in unrelated directions, as a series of numbers rather than a *score*."[19] So tied to the Shakespearean characters and setting are songs like "I've Come to Wive It Wealthily in Padua" and "Where Is the Life That Late I Led?" that "it is difficult to imagine them being sung out of context." The bard's bawdiness licensed Porter's risqué cataloging in such songs as "Brush Up Your Shakespeare," where his list consisted of allusions to various plays:

> If she then wants an all-by-herself night,
> let her rest ev'ry 'leventh or "Twelfth Night."

Yet Porter's success at "musical integration" only proved again that "the better a score is as theater, the fewer of its songs can be singled out" for independent popularity. One of the few songs from the show to become a hit was "So in Love," whose masochistic climax,

> So taunt me and hurt me,
> deceive me, desert me,

may have had its Shakespearean roots in Helena's plea to Demetrius in *A Midsummer Night's Dream,*

> The more you beat me, I will fawn on you.
> Use me but as your spaniel—spurn me, strike me,
> Neglect me, lose me,[20]

but it nevertheless expressed sentiments similar to those in other songs by Porter, as well as many of Hart's.

Yet no sooner had Porter written his first integrated "book" show that he returned to his old ways for such loosely constructed musicals and films as *Out of This World* (1950), *Can-Can*

(1953), and *Silk Stockings* (1955). Such a reversion did, however, produce some old-fashioned hits, such as "From This Moment On," "I Love Paris," and "All of You." Even in these last songs, however, Porter's stylistic split is as wide as ever. Reminiscent of his great list songs of the 1930s, for example, is "It's All Right With Me," which uses an inventory of dismissive negations to probe beneath the urbane nonchalance of his characteristic lover:

> It's the wrong time and the wrong place,
> though your face is charming, it's the wrong face

As the step-wise melody reaches its highest note, the singer betrays his anguish—"It's not *her* face"—but immediately reverts to sensuous cool with "but such a charming face" and the laconically seductive turn he gives to the most ordinary of catchphrases, "that it's all right with me." These shifts of tone continue in the release, swerving from the inane chit-chat of

> you can't know how happy I am that we met,

to the clinically cool detachment of

> I'm strangely attracted to you

By the end of the song, these tonal shifts reach comically absurd dimensions as the singer, after demeaning his present lover's lips as "the wrong lips," for not being "her lips," confesses, like a pleasantly befuddled connoisseur, that they are, nevertheless, "such tempting lips." Then, just as he seems about to give in to passion, he cooly steps back and offers only an arm's-length proposition—"it's all right with me"—couched in a subordinate clause: "if some night you're free."

If "It's All Right With Me" has the tender wit and understated sensuousness of the great list songs of the 1930s—using the list structure itself to both hold off and indulge feeling—Porter's last hit song, from the film *High Society* (1956), was an utter reversion to his most turgidly sentimental vein. "True

Love" fairly drips with feeling unrelieved by even the faintest flash of wit:

> I give to you and you give to me
> true love, true love.
> So on and on it will always be
> true love, true love.

Even more gratifying to Porter than the tears of Louis B. Mayer must have been the praise this song received from the dean of Tin Pan Alley publishers, Max Dreyfus. "In all my sixty-odd years of publishing," Dreyfus wrote him in 1956, "nothing has given me more personal pleasure and gratification than the extraordinary success of your 'True Love.' It is truly a simple, beautiful, tasteful composition worthy of a Franz Schubert."[21] If the "guardian angel on high" invoked in "True Love" was listening to Porter in 1925, when he virtually offered his soul to write sob-ballads, it was surely a fallen one. In this lyric it had the last, infernal word.

8

Conventional Dithers:
Oscar Hammerstein

*Aside from my shortcomings as a wit and rhymester—or,
perhaps, because of them—my inclinations lead me to a
more primitive type of lyric.*

OSCAR HAMMERSTEIN

After the enormous success of *Oklahoma!* in 1943, Oscar Hammerstein took out a quarter-page ad in *Variety*. Instead of trum-
peting his triumph, however, Hammerstein listed all his turkeys
of recent years, from *Free for All* to *Sunny River;* after each show,
he noted the pitifully small number of performances it had had
before closing; finally, in bold type, he crowed, "I'VE DONE IT
BEFORE AND I CAN DO IT AGAIN!" The ad not only exemplified Hammerstein's characteristic modesty; it reminded theater
buffs that *Oklahoma!* was his first real triumph since the 1927
production of *Show Boat.*

During the intervening years, while Rodgers and Hart, the
Gershwins, and Cole Porter were at their insouciant peak, Hammerstein's career went into eclipse. His resurgence, just as the
era of witty sophistication was waning, signaled not only the
beginning of the "integrated" musical, where songs had to fit
plot and character, but the return as well of the lyrical tradition
he represented. Rooted in European operetta, Hammerstein saw
his tradition as "more primitive" than that of "light verse" writers
like Lorenz Hart,[1] a tradition where the lyricist presented forth-

right sentiments rather than witty understatement, where he crafted euphonious, singable phrases—rich in long vowels and liquid consonants—instead of dazzling imagery and deft rhymes.

Nothing better illustrates Hammerstein's adherence to that tradition than the fact that, in 1924, when Lorenz Hart wrote "Manhattan," Hammerstein had his first major hit songs in the operetta *Rose Marie*. While "Manhattan" suited the wittily sophisticated revue *Garrick's Gaieties*, *Rose Marie* followed composer Rudolf Friml's European formula of "old things: a full-blooded libretto with luscious melody, rousing choruses, and romantic passions."[2] Where Hart's language was urbanely vernacular, Hammerstein's soared into the poetic empyrean:

> And yet if I should lose you,
> 'twould mean my very life to me

Here, too, Hammerstein was still following the nineteenth-century practice of dragging out a syllable over three notes ("to-o-o me"), and he dragged them out even more tremulously in "Indian Love Call":

> I am calling you
> -oo-oo-oo-oo-oo

Hart, on the other hand, was deftly using Rodgers' music to split his syllables into clever rhymes and intricate phrases. Not only did Hammerstein work with such European-operetta composers as Rudolf Friml and Sigmund Romberg, he also collaborated with co-lyricist Otto Harbach, who served as Hammerstein's mentor much as Hammerstein himself later tutored young Stephen Sondheim in the art of lyric writing. Hammerstein praised Harbach as one of the "few patient authors who kept on writing well-constructed musical plays" during a period when most theater people would say, "The book of a musical show doesn't matter." "The field of libretto writing," Hammerstein complained, "was therefore filled with hacks and gag men."[3] While few theater historians today would agree with Hammerstein's assessment of Harbach's literary accomplishments, the older lyricist-librettist bequeathed to Hammerstein not only an elevated lyrical style but a desire to anchor lyrics in scene and character.

Hammerstein put that legacy to use in 1927 when he collaborated with Jerome Kern on *Show Boat.* Unlike other shows of the twenties, which Hammerstein characterized as "mostly feather-brained musical comedies with wide-eyed ingenues and earnest young men," *Show Boat* dealt with interracial "and not always happily-ended romance." On the one hand, it looked back to the exotic and grandiose world of operetta, yet it also was a distinctively American period piece. So unlikely a candidate did it seem for a musical of its day that novelist Edna Ferber was at first puzzled why Kern and Hammerstein wanted to use her novel for a show. Given the usual Broadway fare of the day, she naturally thought they envisioned a Ziegfeld girlie spectacle rather than a new kind of American musical drama. While today *Show Boat's* story may seem "melodramatic and gauche," it is still recognized as "the first step toward integrating songs with story in a musical." Because "Hammerstein's lyrics refer to the story and so are sung in character," they force the audience *"to keep the story in mind while enjoying the musical numbers."*[4]

What Ferber's book provided for Hammerstein was the chance to work in a vernacular style as well as an elevated one—writing formal "poetic" songs for white characters but also colloquial, but equally poetic, lyrics for blacks. In a sense, his lyrics for *Show Boat* recapitulate the early history of Tin Pan Alley—blacks declaring their love in slang, as they had in the early "coon" songs, while whites proffered their romantic sentiments in the formal idiom of "After the Ball." ("After the Ball," in fact, was interpolated into *Show Boat* for period flavor.)

In "Why Do I Love You?", for example, the white lovers, Magnolia Hawks and Gaylord Ravenal, ponderously wonder "Why should there be two happy as we?" and "Can you see the why or wherefore I should be the one you care for?" At certain moments, though, Hammerstein gives their lyrics a vernacular touch, as in the hesitant, conversational sequence that was the germ for "Make Believe": "Couldn't you . . . couldn't I . . .

couldn't we . . . ?" No sooner does Hammerstein drop into such colloquial idioms, however, than he soars off again into the florid world of operetta where "our lips are blending in a phantom kiss." Similarly, while he might start off a song with a casually Hart-like "here in my arms," he winds up with "You are spring, bud of romance unfurl'd" (whereas Hart had taken the same phrase, "here in my arms," but followed it with the nonchalant exasperation of "it's adorable—it's deplorable that you were never there!")

When he wrote for the mulatto character Julie La Verne, however, Hammerstein could completely recast the sonorous question "Why Do I Love You?" as a torch song that laconically laments, "Can't Help Lovin' Dat Man."

> Tell me he's lazy,
> tell me he's slow,
> tell me I'm crazy,
> maybe, I know

The off-rhyme, "crazy"/"maybe," marks a perfectly conversational shift, just as another skewed rhyme underscores the singer's sensuously helpless plight: "Ah even loves *him*/when his kisses got *gin*." Compared to that erotic confession, Julie's other *mon homme* number, "Bill," sounds comically genteel. "Bill," with a lyric by P. G. Wodehouse, had been cut from an earlier Kern show but then was interpolated into *Show Boat* (even though the character Julie sings about was named *Steve!*). Where the Julie who sings Hammerstein's "Can't Help Lovin Dat Man" is fully aware that it's gin-soaked kisses that magnetize her, the Julie who sings "Bill" sounds like a slightly daft virgin when she demurely wonders, "I can't explain—it's surely not his brain!—that makes me thrill."

The great hit from *Show Boat,* of course, was "Ol' Man River," a thoroughly theatrical song that nevertheless managed to become independently popular even though it is so clearly tied to a specific character and dramatic situation. Proud of the sense of character and setting he was able to imbue into this lyric,

Hammerstein offered it as an example of how his lyrical skills differed from those of Hart. Asked if he ever used a rhyming dictionary, Hammerstein freely admitted, "I do," but then added, "If you would achieve the rhyming grace and facility of W. S. Gilbert or Lorenz Hart, my advice would be never to open a rhyming dictionary. Don't even own one." Hammerstein then went on to offer a larger distinction between his lyrical art and that of Hart's. While noting that he himself might "on occasion, place a timid, encroaching foot on the territory of these two masters," Hammerstein just as freely admitted, "I never carry my invasion very far. I would not stand a chance with either of them in the field of brilliant light verse. I admire them and envy them their fluidity and humor, but I refuse to compete with them."[5]

"Ol' Man River" illustrates Hammerstein's principle that "rhyme should be unassertive, never standing out too noticeably," for "if a listener is made rhyme-conscious, his interest may be diverted from the story of the song." It is only after ten lines of the refrain for "Ol' Man River," he notes, that there is a rhyme— "cotton"/"forgotten"—a slightly off-rhyme at that, like "sumpin' " and "nothin'," which adds to the prosaic character of the lyric. Nor, he points out, is there a rhyme for the title phrase (he does not mention the subtle echoing of "Ol' " in the repeated "*rollin' along*"). Had he rhymed "river" with "quiver" or "shiver," Hammerstein explains, he could not "have commanded the same attention and respect from a listener, nor would a singer be so likely to concentrate on the meanings of the words." Hammerstein's lyric also requires a singer to assume the dramatic role implied by the words: "a rugged and untutored philosopher" who sings "a song of resignation with protest implied."

While pointing out that such a prosaic character would have been diminished by "brilliant and frequent rhyming," Hammerstein says relatively little about the lyrical strategies he uses in place of rhyme, beyond noting his use of repetition and parallel phrasing. One particularly effective device is his manipulation of verbs (a part of speech seldom used for rhyming) to reflect the thematic tension in the song between the singer's physical power and social powerlessness, a tension contrasted, in turn, to the river's power in repose. The verse contrasts two simple verbs—the blacks "work"[6] while the "white folk play"—then further describes the working blacks in participles, "Pulling dose boats . . . gittin' no rest . . . ," then in imperatives:

> Don't look up and don't look down,
> you don't dast make de white boss frown;
> Bend your knees an' bow yo' head,
> an' pull dat rope until yo're dead.

Here Hammerstein does use rhyme to sharpen the contrast between the white boss, who has only to frown, and the blacks, who must then bend, bow, and pull—but dare not even look. At the end of the verse, the verbs only weakly implore: "Let me go 'way" and "Show me dat stream."

In the chorus—surprisingly a standard Alley thirty-two-bar AABA chorus—Hammerstein turns away from the frenetic world of blacks and whites to the river itself. Along with the shift away from rhyme, the verbs that characterize the river are calm—"must know," "just keeps," even inactive—"don't say," "don't plant." When the lyric shifts back to the human world, strong verbs and sharp rhymes return:

> You an' me, we sweat an' strain,
> body all achin' an' racked wid pain.
> "Tote dat barge!" "Lift dat bale,"
> Git a little drunk an' you'll land in jail

Whereas black and white were juxtaposed in the verse, here they are conjoined in contrast to the serene power of the river, both in their frenzied work and their frustrated ease. With the final A section Kern's melody soars again, and Hammerstein follows it with a new verb form—a sequence of gerunds that portray humans "tryin'," "livin'," and "dyin'," while the river, characterized by a perpetually present participle, "keeps rollin' along."

Show Boat, rather than *Oklahoma!*, might have marked the advent of the "integrated" musical. In fact, had Kern not steadfastly refused to advance along the road opened up by *Show Boat*, he and Hammerstein might well have written *Oklahoma!*—a dozen years ahead of its time. In 1931 Hammerstein tried to persuade Kern to make a musical out of a regional play by Rolla Lynn Riggs, *Green Grow the Lilacs*, the book on which *Oklahoma!* later was based. Ever since 1920, however, Kern had been drifting away from the innovative style of his Princess shows toward more traditional—and lavish—musicals such as *Sally* (1920), and its clone, *Sunny* (1925). After *Show Boat*, he continued that reversion, with *Sweet Adeline* (1929), and, for the next decade, had his greatest successes on Broadway with shows that alternated between grandiose operetta and frothy musicals. These shows, such as *Roberta* (1933), concentrated on the song rather than the book, and it was usually Otto Harbach, rather than Hammerstein, who supplied high-pitched lyrics for soaring melodies like "Yesterdays":

> Then gay Youth was mine,
> Truth was mine,
> joyous free and flaming life
> forsooth was mine

Equally sententious was "Smoke Gets in Your Eyes," where Harbach threw in the archaic kitchen sink with "So I chaffed them and I gayly laughed" (a line that seems especially glaring in the rock version of the song done by the Platters in the 1950s).

Whatever hopes Hammerstein still nurtured for writing musical *drama*, rather than musical comedies that merely showcased songs aimed at Tin Pan Alley's pop market, became even dimmer when he followed the general exodus of songwriters to Hollywood. Hollywood did not want songs grounded in character or dramatic situation, and of all the Broadway refugees on the West Coast, none was more disgruntled than Hammerstein. Occasionally the studios assigned him to a picture that suited his lyrical style, such as the 1937 film *High, Wide, and Handsome*, which was set in the late nineteenth century and thus called for period songs. In typical Hollywood fashion, however, characters could not simply burst into song the way they did on Broadway; they had to have an "excuse" for singing. In *High, Wide, and Handsome,* Irene Dunne played the part of a singer—married to Randolph Scott—and thus could "perform" her songs in saloons and carnivals.

The hit of the film was "Can I Forget You," with its long, nostalgic lines—"once we walked in a moonlit dream . . . how sweet you made the moonlight seem." In the release both Kern and Hammerstein pull out all stops—then just as suddenly mute them:

> Will the glory
> of your nearness
> fade—
> as moonlight fades in a veil of rain

The sudden fading into the delicate image of a veil of rain compensates for the stentorian vagueness of "glory of your nearness."

Hammerstein, however, saw his artistry not in creating such imagery but in what he called "phonetics"—the careful manipulation of vowel and consonant sounds (in this case the repetition of the same open *a* vowel through "f*a*des in a v*ei*l of r*ai*n). Such phonetics, he insisted, make lyrics eminently "singable." "Wherever there are vocal climaxes and high notes," he pointed out, "singers are comfortable only with vowels of an open sound."[7] Such "phonetics" characterize the few hit songs Hammerstein wrote during the 1930s, like "The Song Is You" (1932) and "All the Things You Are" (1939). While Hart, Gershwin, and Porter flashed brilliant imagery and reveled in clever rhyme, Hammerstein stuck to the traditional lyricist's task of finding euphonious words that effaced themselves before Kern's lush music.

Despite such occasional hits, right down to "The Last Time I Saw Paris" (1940), by the early 1940s Hammerstein had acquired a string of flop shows and forgettable movies. Turning his back on both Hollywood and Broadway, he retreated to his Pennsylvania farm and began a labor of love—writing a libretto for an

Americanized version of *Carmen*. It was while he was at work on *Carmen Jones* that Richard Rodgers approached him about collaborating on a musical based on Rolla Lynn Riggs' play, *Green Grow the Lilacs*. Instead of jumping at the chance to do something he had wanted to do for years, Hammerstein gallantly insisted that Rodgers continue to try to work with his longtime partner. If Hart could not complete the lyrics, Hammerstein offered to step in secretly and help out, receiving no credit for himself. Only when Hart flatly refused to work on *Green Grow the Lilacs*, telling Rodgers such a corny story could never be a successful musical, did Hammerstein agree to the collaboration. When Rodgers broke the news to Hart, whose lyrical style was the antithesis of Hammerstein's, Hart said "You couldn't have picked a better lyricist."

What Rodgers realized instinctively was that the sophistication of Rodgers and Hart, like that of Porter and the Gershwins, was becoming passé by the early years of World War II. By blending his music with Hammerstein's older lyrical style, he hoped, something new would emerge. In turning to *Green Grow the Lilacs*, moreover, Rodgers, as we have seen, was following the shift in all the arts—from Benton's murals to Steinbeck's novels—away from urbanity toward regionalism, folk simplicity, an emphasis upon the local and communal. For Hammerstein, that shift gave him the same opportunity he had had in *Show Boat*—that of grounding his highflown sentiments in particular characters and situations. In *Oklahoma!*, however, all of those characters were regional folk whose lyrics called for a vernacular, rather than a "poetic," operetta style.

The shift was apparent from their very first song—in fact, even before they wrote the first song. "The traditions of musical comedy," Hammerstein recalled, "demand that not too long after the rise of the curtain the audience should be treated to one of musical comedy's most attractive assets—the sight of pretty girls in pretty clothes moving about the stage, the sound of their vital young voices supporting the principals in their songs." Torn between "showmanship" and "dramaturgy," Rodgers and Hammerstein decided to break with tradition and present the "story in the real and natural way."[8] The curtain would open on a solitary farm woman churning butter.

The opening song, sung offstage by Curly, was "Oh, What a Beautiful Mornin'." It was also the first collaboration of the new team of Rodgers and Hammerstein and revealed what would be different about musicals from then on. Like most of their songs, the words came first—a reversal, as Hammerstein himself noted, of Tin Pan Alley priorities (though not enough of a reversal to violate Ira Gershwin's axiom that the art of lyric writing consists of fitting words "mosaically" to music already composed; Hammerstein wrote his lyrics with snatches of other music—not Rodgers'—in mind). Still, for Rodgers the difference was profound. With Hart he had been a "popular" songwriter—"writing tunes and leaving it to the lyricist to set specific thoughts to." But with Hammerstein, he "tried to match his music to the lyricist's meaning, feelings, forms, and theatrical purposes."[9]

Those lyrical meanings, moreover, emerged from the libretto. The opening stage directions for *Green Grow the Lilacs* intrigued Hammerstein:

> It is a radiant summer morning several years ago, the kind of morning which, enveloping the shapes of earth—men, cattle in the meadow, blades of the young corn, streams—makes them seem to exist now for the first time, their images giving off a visible golden emanation that is partly true and partly a trick of imagination, focusing to keep alive a loveliness that may pass away.[10]

In adapting these images to set the tone of the play, Hammerstein might have soared into the sonorities of "Can I Forget You?" or "All the Things You Are." What saved him was his recognition that he had to bring "the words down to the more primitive poetic level of Curly's character," so the "visible golden emanation" becomes "a bright golden haze on the meader'."

Along with integrating his lyrics into the regional story and character, Hammerstein maintained his commitment to "phonetics"; the long vowels of "Oh, What a Beautiful Mornin' " made it eminently singable, he noted, and the clipped "mornin' " softens the final consonant at the same time that it gives the sentiment a colloquial turn. At times, the conflicting pulls of semantics and phonetics produced problems. Hammerstein had originally had Curly compare the growing corn to "a cow-pony's eye," but he noticed that his neighbor's field of corn was already considerably higher than any pony. The phrase was also too consonant-laden for a singer to enunciate clearly, so Hammerstein reluctantly substituted "elephant's eye." The new image was not only more accurate and singable but more in character; a farm-boy like Curly is more likely to reach for his metaphors in the memory of a rare glimpse of elephants in a traveling circus, rather than in the common sight of the ordinary denizens of his barnyard (equally apt is his citified simile "The cattle are standin' like statues"). Hammerstein's skill with phonetics stretches across all the songs in the show: the long vowels of the show's opening phrases,

> Oh! what a beautiful mornin'!
> Oh what a beautiful day

arc recapitulated in the final phrase "Oklahoma—*O-K*," and reverberate through such varied lines as "I Cain't S*a*y N*o*," "Don't thr*ow* b*ou*q*ue*ts at me," and

> Everythings's up-to-date in Kansas City,
> Th*ey*'ve gone about as f*u*r as th*ey* c'n g*o*!

Not only did Hammerstein integrate his lyrics into dramatic contexts, he was frequently able, as in "Ol' Man River," to integrate dramatic action, setting, and character into his lyrics. Even in such a straightforward love song as "People Will Say We're in Love," Hammerstein adds bits of dramatic stage business between Curly and Laurey:

> Don't start collecting things—
> give me my rose and my glove!

So prominent is character and action here that it is difficult to recognize that "People Will Say We're in Love" is a list song in the tradition of Cole Porter. While Porter's images flash out at the listener, however, Hammerstein's inventory is barely noticeable as a catalog, so that we concentrate not on watching the items "top" one another but on the homespun romantic situation. So daringly does Hammerstein subvert this most Porterish of genres, he even makes listing the subject of the song, announcing in the verse that this is to be a song about "linking up" things, a "practical list of 'don'ts.'" Similarly, where Porter flaunts his intricate rhymes, Hammerstein buries them, so that one barely hears "wish" resonate with "carved your in*it*ials on the tree."

Hammerstein keeps character and scene prominent even when his imagery is at its flashiest:

> The wheels are yeller, the upholstery's brown,
> the dashboards's genuine leather,
> with isinglass curtains y' c'n roll right down—
> in case there's a change in the weather.

Again, even the cleverest rhymes don't upstage Curly and his deadpan tall-talk:

> "Would y' say the fringe was made of silk?"
> Wouldn't have no other kind but silk.
> "Has it really got a team of snow-white horses?"
> One's like snow—the other's more like milk.

The ever-so-faint off-rhyme *silk/milk* cleverly matches the slightly mismatched horses, yet Hammerstein's off-handed dexterity of rhyme, as always, plays second fiddle to his dramatic sense of character—in this case, the Yankee peddlar streak in his Oklahoma cowboy.

Oklahoma!, therefore, provided theater lyricists with a new idiom; neither the florid, elevated style of European operetta, nor the sophisticated urbanity of Hart, Gershwin, and Porter, but a style that took its cue from the regional character who voiced it.

For Hammerstein, that style would permit sentiments as "corny as Kansas in August," and for lyricists who followed him it lent itself to characters as diverse as New York street gangs and citizens of River City, Iowa. *Oklahoma!* thus gave the musical a "shape that was basically American":

> it told a story of reasonably adult interest and did not suffer the story to be intensified by irrelevant songs, dances, ballets and bursts of comic patter. On the contrary. The authors had aimed at making every song, dance, ballet and joke a means of advancing the story.[11]

In doing so, *Oklahoma!* represented the triumph of the book over the free-standing lyric, the triumph of the Broadway musical as *theater*, not as a supermarket for the wares of Tin Pan Alley.

The completeness of that triumph is evident today, when even the most successful Broadway musical almost never produces a single popular song—a far cry from the golden age, when a hit show was, simply, a show with a lot of hit songs. Yet even the staunchest advocates of Hammerstein's central role in redefining theater lyrics cast a furtive look back at the lyrics of the golden age. Noting that "writing to character and plot means giving up wit, unless your show happens to be about some smart set," Martin Gottfried admits "sophisticated and literate lyrics are among the great joys of musical comedy. Would we even have *wanted* Hart or Porter to write for character and deny their wit?"[12] So, too, Gerald Mast, while praising Oscar Hammerstein as "the premier poet of the American musical theater" who "redefined the singer of a song" as a "specific character living in a specific place at a specific moment in history"—an Oklahoma rancher, or a nurse from Little Rock, or a Victorian British schoolmarm—he, too, laments the loss of "the great song hits from Broadway shows" made possible by the fact that "the voice, the I, of a Gershwin, Hart, or Porter lyric is an undefined surrogate for the lyricist himself."[13]

That urbane voice, as we have seen, was easily tailored to Tin Pan Alley's commercial market, where, for a generation it set a new standard in popular song. Given the fact that so many of those "unintegrated" theater songs have endured as "standards" while the shows that spawned them have long been forgotten, the victory of the Broadway musical over Tin Pan Alley may well have been a Pyrrhic one.

Paper Moons:
Howard Dietz and
Yip Harburg

I doubt that I can ever say 'I Love You' head on—it's not the way I think. For me the task is never to say the thing directly, and yet to say it—to think in a curve, so to speak.

E. Y. "YIP" HARBURG

In his flamboyant autobiography, *Dancing in the Dark,* Howard Dietz recounts how, sitting in a New York bar one day in 1929, he overheard two producers discussing plans for a new revue.

> *The Little Show* was to be a revue, but not in any respect like the rhinestone creations with huge staircases of Flo Ziegfeld or Earl Carroll, the G-string titivator. If it was to be compared to any show, it got its inspiration from *The Garrick Gaieties*. It was to be topical and artistic, a witty travesty in the leitmotif, if possible.[1]

Dietz "careened over to their table," recited one of his own poems—about show business flops—and was teamed with Arthur Schwartz to write the songs.

The Little Show inaugurated a string of smart and intimate revues, sometimes laced with political satire, tailored to a frugal, Depression-era Broadway that flourished during the 1930s. These revues brought new lyricists to the fore, such as Dietz, who wrote songs with Arthur Schwartz for *The Band Wagon* (1931), *Revenge With Music* (1934), and *Between the Devil* (1937), and E. Y. "Yip" Harburg, who collaborated with various composers on similar shows, such as *Americana* (1932), *Life Begins at 8:40* (1934), and *Hooray for What* (1937).

In these revues the emphasis was not upon lavish sets and sumptuous spectacle but on satirical sketches and sophisticated songs. Given that emphasis, it is striking that, while both lyricists could occasionally write in the urbanely witty style of the golden age—it's been argued, for example, that "as Harburg was influenced by Ira Gershwin, Howard Dietz was influenced by Hart"[2]— they more often opted for Hammerstein's mode of straightforward sentiment underscored by subtle "phonetics." While Hammerstein's adherence to that elevated style reflected his admitted "shortcomings as a wit and rhymester" in the "field of brilliant light verse," Dietz and Harburg's commitment is more puzzling. Both men had started out as writers of *vers de société*, and their poems appeared frequently in F. P. A.'s "Conning Tower." While they carried that light verse style over into their comedy songs, however, for romantic ballads they usually reverted to the traditional lyrical style of elevated diction and highflown sentiment.

Part of that reversion may reflect the fact that their collaborators frequently supplied them with melodies that emulated the lush sonority of Jerome Kern, who, as we have seen, could make any lyricist, even Ira Gershwin, turn purple. It is more likely that the reversion stemmed from Dietz and Harburg's sense of themselves as "theater" writers, whose songs called for a more poetic idiom, a cut above the Alley's more slangy fare. Well before the advent of the integrated musical, they saw themselves not as songwriters who "provided songs for the stage" (like Cole Porter) but rather as "theater" writers who also wrote songs.[3] Yet while both Dietz and Harburg conceived of their lyrics in theatrical terms, their sense of what constituted a "theater" song differed markedly. For Dietz, a "theater" song was one designed to be presented on stage, replete with costume, lighting, sets, and choreography. When Harburg wrote a "theater" song, however, he thought in terms of writing lyrics for a particular character in a certain dramatic situation. Thus Harburg was able to adapt to the new integrated musical of the 1940s, while Dietz's career as a lyricist was largely confined to the stage revues of the 1930s.

Long before he had established himself as a successful lyricist, Dietz was well known as one of several self-styled "literary aficionados," who "lived by a code of scansion" and frequently placed their poems, alongside those of Dorothy Parker and Robert Benchley, in F. P. A.'s "Conning Tower." In his first hit song, "I Guess I'll Have to Change My Plan" (1929), he proved that he could, when his sense of "theater" called for it, turn his society verse talents into an urbanely colloquial lyric. Clifton Webb, the debonair star of *The Little Show*, "wanted a number that was more perverse, a number he could deliver all alone in full-dress suit and a spotlight . . . a lyric with suave romantic frustration."[4]

Composer Arthur Schwartz tried to accommodate Webb with what seems an unlikely tactic—digging up "I Love to Lie Awake in Bed," a song he had written years before with Lorenz Hart when both boys worked at a summer camp. Hart's original lyric had dripped with adolescent gush:

> I rest my head upon my pillow—
> Oh, what a light the moonbeams shed.
> I feel so happy I could cry,
> and tears are born
> within the corn-
> er of my eye

But Dietz turned this weepy camper into a jaded roué who, at first, nonchalantly laments the fact that he must "change his plan," since the object of his affection turns out to be married:

> Why did I buy those blue pajamas
> before the big affair began?

Then Dietz gave Webb precisely the "perverse" twist he wanted, by having this sophisticate casually change his plan again—to adultery—and even relish the added delight:

> Forbidden fruit I've heard is better to taste.
> Why should I let this go to waste?

That resolve gives a clever turn to the closing phrase, transforming the woebegone "I've lost the one girl I found" to the worldly "I've found the one girl I've lost."

Dietz also showed he could rhyme as deftly as Hart, deflating "bliss is" with "Mrs." and lightening the agony of "dwelling in my personal hell." Equally evident is his skill in vernacular phrasing. Lehman Engel has pointed out that the "levity of this lyric is established in a single word in the title: guess."[5] Dietz enhances the colloquial ease with such nonchalant phrases as "I overlooked that point completely" and "I should have realized," followed by the slangy contraction "there'd be another man." Not only can Dietz use contractions as skillfully as Gershwin, he makes a simple rhyme like "that/at" carry understated emotional weight:

> Before I knew where I was at
> I found myself
> upon the shelf,
> and that was that

He gives the insignificant "that" even more power by carrying it over into the next lines where it plays off "there," just as "get" ironically undercuts "got":

> I tried to reach the moon but when I got there,
> all that I could get was the air

Such slangy circumlocution neatly underscores the futile romantic effort that goes nowhere, gets nothing.

"I Guess I'll Have to Change My Plan" epitomizes the style of the golden age—a witty, vernacular lyric that vies with Hart, Gershwin, and Porter at their best. Yet Dietz seldom used this style again in his romantic songs. Instead he opted for florid melodrama, reserving his clever rhymes and literate wit for comic production numbers. At times, the choice may have been dictated by Arthur Schwartz's melodies, which, as Alec Wilder points out, often resemble Jerome Kern's and seem to call for an equally sumptuous lyric. Yet Dietz himself tells the story of how he induced his collaborator to change an upbeat melody into a lugubrious one. Schwartz had written a fast song called "I Have No Words," with a lyric by British songwriter Desmond Carter. Dietz suggested he play it in "ballad tempo," but "every time he repeated it, he seemed to play it at a gallop." Stubbornly insisting that the composer "slow down to the pace I had in mind," Dietz then set new words to the melody. While not especially clever, Carter's original lyric for "I Have No Words" was clearly in the Hart-Gershwin-Porter vein—nonchalantly sophisticated, flippantly unsentimental:

> I would beg for you,
> break a leg for you,
> lay an egg for you

When Dietz rewrote the song as "Something to Remember You By," however, he began with a melodramatic image he would use again and again in his lyrics: "You are leaving me and I will try to face the world alone." From there he slides into a bit of stoicism—"what will be will be"—countered by a lump of nostalgia: "but time cannot erase the love we've known." As the verse draws to a close, it oozes with poetic diction:

> Let me but have a token
> thru which your love is spoken

In the chorus the sentimentality gets so thick it invites risqué parody, as the lover pleads for "some little something, meaning love cannot die, no matter where you chance to be." Compared with Ira Gershwin's handling of the same theme in "They Can't Take That Away From Me," with its antiromantic "the way you sing off key," Dietz's lyric seems an anachronism in the golden age.

Yet Dietz defended his traditional lyrical practice. It was he, after all, who criticized Lorenz Hart's pyrotechnic rhyming, arguing that a good lyric does not depend upon rhymes like "choose a sweet lollapaloosa" but on carefully manipulating vowels, "lots of *a*'s and *o*'s and *I*'s and *ah*'s—even *e* words are not openly singable." Dietz's substitution of "pray for you . . . night and day for you . . . see me through" for Carter's original sequence, "beg for you . . . break a leg for you . . . lay an egg for you," was not only a substitution of elevated diction for witty slang; it also dropped Carter's clever rhymes for long vowels that make the lyric more sonorously singable.

Dietz here clearly aligns himself with Hammerstein's emphasis upon "phonetics," though it was Irving Berlin whom Dietz praised as a lyricist "who counts the vowels better than most of his contemporaries." Even more indicative of Dietz's traditional stand was his attitude toward Cole Porter. On the one hand, he praised Porter's simple, sentimental lyrics that lovers "kissed to" when "lights were low." But he mocked Porter's brilliant list songs as mere exercises in rhyming "a Sears and Roebuck catalogue." Yet another indication of Dietz's reactionary stance was his choice of Maxwell Anderson and Kurt Weill's "September Song" as his favorite lyric. The antithesis of casual sophistication, "September Song" (1938) must have appealed to Dietz not only for its "long, long" vowels, but because it epitomizes the kind of "theatrical" song he aspired to write. It was written, in fact, not for a singer but an actor, Walter Huston in *Knickerbocker Holiday*, who delivered it more as a dramatic reading, his voice thick with melodramatic nostalgia.

It was just such lyrical *Sturm und Drang* Dietz emulated, and in Arthur Schwartz he found a collaborator who could supply melodies with the same brooding theatricality, like their big hit from *The Band Wagon*, "Dancing in the Dark" (1931). What Alec Wilder jokingly calls a "Mata Hari song,"[6] "Dancing in the Dark" is based on the musical pattern of returning, insistently, to the same phrase, repeated at step-wise intervals. Those repetitive musical revolutions inspired Dietz with the idea of using a revolving stage when the song was performed; to cap the "revolving" effect Dietz had the stage surrounded with mirrors so that Tilly Losch danced with multiple images of herself while John Barker sang.

Dietz then concocted a "revolving" lyric that relied upon parallel phrases, phrases made even more repetitive by alliteration: "dancing in the dark," "waltzing in the wonder," "looking for the light." As in "Something to Remember You By," he uses few rhymes—none whatsoever for the title phrase—and relies upon clever enjambment to get a "whirl-and-step" effect as one phrase suddenly pauses, then melodramatically continues:

> time hurries by—
> we're here
> and gone

Such ten-cent existentialism stays at a high level of diction until the end of the chorus, when a vernacular catch-phrase suddenly intrudes amid the general bombast: "And we can face the music—together!"

In "Alone Together" (1932) Schwartz provided a melody that, once again, climbs higher and higher in step-wise repetitions, and Dietz follows them with parallel phrases—"beyond the crowd, above the world"—that become increasingly bathetic:

> The blinding rain,
> the starless night,
> were not in vain

When a colloquial idiom—like "we can weather"—does get in, it is quickly followed by a metaphysical leap to "the Great Unknown"—hardly a thing one "weathers," a stylistic clash made all the more wrenching since it is Dietz's sole rhyme for his oxymoronic title phrase—"Al*one* (Unk*nown*) tog*ether* (we*ather*)."

When Schwartz hatched another brooding dance in the dark, "You and the Night and the Music" (1934), Dietz again followed with swirling alliterative phrases—"fill me with flaming desire"—now made even more tempestuous with internal rhymes—"thrill me but will we be one?"—that even violate good grammar: "after the night and the music are DONE!" Even when Schwartz and Dietz start off with a nonchalant musical and lyrical line, as in the title phrase of "If There Is Someone Lovelier Than You," (1934), they suddenly shift into melodramatic declamation,

> then I am blind,
> a man without a mind

raise the stentorian pitch higher,

> But no, I am not blind,
> my eyes have travelled ev'rywhere,

and climax with an operatic oath: "By all that's beautiful, such beauty can't be true."

It was not until their last revue of the 1930s, *Between the Devil* (1937), that Dietz and Schwartz returned to the urbane style of "I Guess I'll Have to Change My Plan." Perhaps inspired by the fact that the song was written for Fred Astaire, the writers came up with a jauntily heartbroken lament, "By Myself":

> The party's over, the game is ended,
> the dreams I dreamed went up in smoke,
> they didn't pan out as intended;
> I should know how to take a joke.

As in "I Guess I'll Have to Change My Plan," Dietz structures the chorus around a series of parallel vernacular phrases: "I'll go my way . . . ," "I'll have to . . . ," "I'll try to. . . ." Within this sequence of casual contractions even Dietz's favorite image of the "facing the Great Unknown" gets a light touch: "I'll face the unknown." Instead of building up bombastic declamations, the lyric turns a witty play on the preposition "by," so that "by myself" means both "I'm alone" as well as "I did it without anybody's help":

> No one knows better than I myself
> I'm by myself alone

That turn lets Dietz sidestep sentimentality by celebrating self-sufficiency at the same time that he laments loneliness.

Dietz and Schwartz went their separate ways, too, after 1937, as Dietz found his Hollywood duties as publicity director for MGM demanding more and more of his time. They did collaborate occasionally again—in 1948 for another revue, *Inside U.S.A.,* then in 1953 for the movie version of *The Band Wagon* (for which they added the brilliant list song, "That's Entertainment"), and even on two unsuccessful musicals in the 1960s. Their songs, however, remain tied to the topical revues of the 1930s. While Schwartz later went on to write such integrated musicals as *A Tree Grows in Brooklyn,* with librettist-lyricist Dorothy Fields, Dietz's career as a lyricist came to an end with the passing of the revue. As Martin Gottfried concludes, he "had little interest in relevant, plot-justified, or character lyrics," and "when the theater moved away from revues and toward book musicals" he was left behind.[7]

Like Dietz, E. Y. "Yip" Harburg established himself as a lyricist in revues of the 1930s, though from the first he manifested a more complex "theatricality," one that depended less on spectacle than on having a lyric grow out of character and situation. Harburg, too, had his roots in *vers de société,* roots that went back, as we have seen, to his high-school friendship with Ira Gershwin. During the 1920s Harburg occasionally published a witty poem in F. P. A.'s "Conning Tower," or contributed a song to a musical, but his main interest lay in business, and by the end of the decade he had built up a successful appliance firm. It was only when that business collapsed in the Depression, that Harburg, with his friend Ira's help, turned to songwriting. "I had my fill," he observed wryly, "of this dreamy abstract thing called business and I decided to face reality by writing lyrics."[8]

Teamed with composer Jay Gorney in the satirical revue *Americana* (1932), Harburg wrote one of the most unusual hits of the golden age, "Brother, Can You Spare a Dime?," the song most often cited as an exception to the rule that all popular songs from this era were about love. Harburg's description of the genesis of the song indicates that, even more than for Hammerstein or Dietz, a lyric had to be grounded in a particular character and dramatic situation. To keep that ground clear, Harburg relied upon simple, straightforward lyrical techniques, rarely indulging in the dazzling rhymes and wordplay of Hart or Gershwin unless he was writing a comic number.

In "Brother, Can You Spare a Dime?," Harburg, deeply political himself, took Roosevelt's campaign image of the "Forgotten Man" and created a dramatic monologue rooted in character:

> The fellow in the breadline, just back from the wars . . . a bewildered hero with a medal on his chest ignominiously dumped into a breadline. I wanted a song that would express his indignation over having worked hard in the system only to be discarded when the system had no use for him. . . . Thus: "Once I built a railroad, made it run, made it race against time. . . ." He's still feeling his strength, and brings that strength into the song. But suddenly he looks at himself and stops short, puzzled: "How the hell did I get into this position, where I find myself saying, "Brother, Can You Spare a Dime?"[9]

The monologue is framed by the catch-phrase title, transformed from a panhandler's pathetic plea to a veiled threat. The bitterness is quietly planted at the opening of the chorus, as the singer recalls how "they used to tell me I was building a dream" and reminisces romantically about going off to war:

> Once in khaki suits
> Gee, we looked swell
> full of that Yankee Doodle-de-dum

But the reminiscence quickly sours, as the "half a million boots" that jauntily march "Over There" go "sloggin' thru Hell." That image carries a further implication—that the same soldiers might now band together in revolutionary protest, an implication that hangs fire between the powerful, active verbs that recount the past—"built a tower," "made it run," "went slogging through hell"—and the participles "standing" and "waiting" that uneasily mark time for the present. The suggestion of revolution grows stronger with the reference to "Yankee-Doodle-de-dum" and the allusion to the "Spirit of '76"—"I was the kid with the drum." Such touches give the singer a mythic power as he dramatically steps closer to his "brother":

> Say don't you remember,
> they called me Al
> it was Al all the time.
> Say, don't you remember
> I'm your pal!

The rhymes and near-rhymes—Al/all/pal—give this intimate identification an aggressive edge, capped in the last line when "brother" is suddenly replaced by the military—and militant— "Buddy, can you spare a dime?"

Despite its political emphasis, "Brother, Can You Spare a Dime?" has some of the lineaments of a standard Alley love song. The plight of the ex-soldier, for example, parallels countless romantic laments where a jilted lover bemoans the girl that did him wrong. Harburg himself noted that what he wanted to avoid in the lyric was "sentimental, tear-jerking . . . maudlinity"—precisely what Hart and Gershwin strove to avoid in their love songs. What Harburg used to avoid those pitfalls, however, was not flippant rhymes and witty word-play but a realistic portrayal of character and dramatic situation.

He could use the same devices in a straightforward love song, as when he set a brooding, nostalgic lyric to Vernon Duke's "April in Paris" (1932). It was theatrical necessity that mothered his invention: the producer of the revue *Walk a Little Faster* had ordered a Parisian set and told Harburg to write a song to go with it. Harburg, who had never been to Paris, read up on travel brochures to create a lyrical setting, and then came up with an unusual character to place in it. Instead of the clichéd lover sitting in Paris and reminiscing about an old flame, Harburg takes a singer who has never been in love but, in an even greater tribute to Paris' power in April, wishes he had been—so that now he *could* have romantic memories.

Duke's melody, according to Harburg, was "light and airy,

and very smart," steeped in cosmopolitan sophistication—"all of that Noel Coward/Diaghilev/Paris/Russia background." Neverthe-less, Harburg added, in a comment that indicates his commit-ment to the elevated and sentimental tradition, it lacked "the essential theatricalism and the histrionics that writing for shows demanded—the drama, the emotions."[10] It was those "theatrical" qualities that Harburg reached for in his lyrics for "April in Paris," rather than a witty elegance that would parallel Duke's music. As a result, the song was a "strange mixture" between sophistication and sentimentality, a mixture like that of a Rod-gers and Hart song, but with Harburg and Duke the formula was reversed—melodramatic lyrics and urbane music.

Instead of the flippant imagery Hart would have concocted to debunk the sonority of a Rodgers melody, Harburg came up with unobtrusive clichés that effaced themselves before the ele-gance of Duke's music:

> I never knew the charm of spring . . .
> I never knew my heart could sing . . .
> never missed a warm embrace . . .

Similarly, rather than Hart's vernacular idiom, Harburg opted for such archly poetic phrases as "whom can I run to?" and "a feeling no one can ever reprise." Duke's refrain uses a casual rhythmic phrase—a triplet followed by quarter- and whole notes; a less skillful craftsman might have tried to fit the triplet to a three-syllable word, but Harburg deftly plays lyrical against musi-cal phrases, setting the triplet tremulously to two words, "*Ap*-ril-in" and *chest*-nuts-in," which form dactylic feet, thus creating a more dramatic emphasis on the follow-up trochees, "*Pa*ris" and "*blos*som."

There are few rhymes in "April in Paris," and none of them calls attention to itself the way Hart's rhymes do. Harburg opts instead for off-rhymes, "*warm*" and "*charm*," and waits until the end of the release—with the lingering "till" to rhyme his title, as the last A section begins "Ap*ril* in Paris." "Par*is*," too, finally receives even fainter rhymes: "th*is is* a feeling." Instead of

rhyme, Harburg relies more on Hammerstein's "phonetics," the
long *a* of "*A*pril," hauntingly resurfacing in "holid*a*y *ta*bles," then
again under the *face/embrace* rhyme. Like Dietz and Hammerstein, Harburg sensed that pyrotechnic rhyming would disrupt
the melancholy mood of the lyric. It is interesting to compare his
muted rhymes to those in "Autumn in New York" (1934), a
follow-up to "April in Paris." This time Vernon Duke wrote his
own lyric, and used numerous clever rhymes,

> glittering crowds and shimmering clouds . . .
> jaded roués and gay divorcees . . .

giving the lyric a jaunty urbanity that clashes with his efforts to
create a melancholy mood of joy "often mingled in pain."

Even in his most straightforward romantic hits, such as "It's
Only a Paper Moon" (1933), Harburg could take the frothiest
Alley clichés and give them a twist to fit the dramatic context. *The
Great Magoo*, with a book by Ben Hecht and Charles MacArthur,
called for a love song for a "very cynical" Coney Island barker.
Trying to think of a cynical angle on love, Harburg took an idea he
would use again in "Fun to Be Fooled" (1934), creating a theatrical set out of the clichés of Tin Pan Alley lyrics, parodying them to
expose their tawdry artificiality. The oldest claptrap of love songs,
moon and sky, sea and tree, get a cynical and slangy rebirth from
the exposure: "It's only a paper moon, sailing over a cardboard
sea," and "only a canvas sky, hanging over a muslin tree." The
cynicism is made all the sharper—and the tired images made
more vivid—by specifying their cheap, manufactured, materials:
paper, cardboard, canvas, muslin. Not only is it all fake, it's cheap
fake, from a "honky-tonk parade," through "penny arcade," and
even fake that everyone knows is a fake: "It's a Barnum and Bailey
world." Thus the straightforward romantic plea,

> But it wouldn't be make believe
> if you believed in me

comes off as just more Tin Pan Alley cliché. Far from suggesting
that love transforms a "world just as phony as it can be" the

formula invites us in, like the pitch of a sideshow barker, where "it's fun to be fooled."

Harburg further drives home the thoroughgoing artificiality of the world of popular song by using one rhyme pervasively through the three A sections—*sea, me, believe, tree, only, be, Bailey*—even carrying it into the release with "hon*ky*," "melo*dy*," and pen*ny*." The only other rhyme is the release's "par*ade*," "pl*ayed*," and ar*cade*." But he brackets those incessant rhymes with harsh consonants that underscore the cheap fabric of "cardboard" and "canvas," "honky-tonk parade" and "penny arcade," "Barnum and Bailey" and "make-believe." With an eye on the "phonetics" of Hammerstein, Harburg makes the most euphonious word in the song the slangy and semantically bitter: "phony as it can be."

Harburg's collaborator in "It's Only a Paper Moon" was Harold Arlen, with whom he had what he termed his "longest and happiest songwriting association."[11] Together they wrote a number of satirical revues, such as *Life Begins at 8:40* (1934, with Ira Gershwin) and *Hooray for What* (1937), but none of their witty songs—not even "What Can You Say in a Love Song (That Hasn't Been Said Before)?"—became popular. Like most songwriters, Arlen and Harburg spent much of the decade in Hollywood, but here, too, few of their songs detached themselves from films to achieve independent success.

Harburg tried to explain that situation by noting that, whether he was writing for the theater or the movies, he was a "pragmatic" lyricist—one who tried to "direct the lyric" to what was happening "on screen or on stage." Thinking in "showmanship terms," working with lyrics "as a director would work, and as a book-writer would," Harburg found it was "pretty hard" to have a song "step out of the histrionic medium and plot—which it accelerates—and be made to flourish and blossom" into a "popular hit." As a consequence, while such witty songs as "Lydia, the Tattooed Lady" (from the 1939 Marx brothers film *A Day at the Circus*), were perfect showpieces, they did not transcend what Harburg called "the pragmatic medium."[12]

The one song that did become not only Harburg's most popular song but one of the most successful songs of all time was written over Harburg's objections to its lack of integration in that pragmatic medium. Chosen to do the songs for *The Wizard of Oz* (the producers had initially wanted Jerome Kern to do the music—and Shirley Temple to play Dorothy!), Harburg and Arlen approached the task in theatrical terms. "I loved the idea," Harburg recalled, "of having the freedom to do lyrics that were not just songs but *scenes*."[13] That phrase, lyrics that were not "just songs," epitomizes the aspiration of a lyricist like Harburg, and in *The Wizard of Oz*, he had the chance to eliminate what he called "stop-plot" songs and write verse dialogue and songs, like those set in Munchkin Land, which stretched across entire scenes.

It was Arlen who felt such "pragmatic" songs had to be balanced with a ballad—"something with sweep, a melody with a broad, long line."[14] When he first played his "lush" melody for Harburg, the lyricist objected on his usual "character" grounds: it sounded like a song for Nelson Eddy—not an eleven-year-old girl from Kansas. It was only when Arlen, at the suggestion of Ira Gershwin, toned down some of the music's grandeur that Harburg agreed to do the lyric for "Over the Rainbow." Proudly noting that Frank Baum's book never mentions a rainbow, he explained that the image came to him as the only truly colorful thing a Kansas farmgirl could use to brighten up her daydreams of a more exotic land. Whatever colorful objections heartlanders might raise at such East Coast/West Coast condescension, the rainbow image certainly suited the idea of doing the opening of the film in black and white, the dream sequence in technicolor.

Thus even this "scene-stopper" was integral to character and setting. For all its independent popularity, the very mention of the title, arching, rainbow-like, between two long vowels—*over the rainbow*—inevitably evokes its dramatic context. Amazingly, the studio cut the song from three successive prints of the film. After each cut, however, associate producer Arthur Freed fought to get it back in, even against the music publishers, who objected that the octave-long interval between "some-" and "-where" made it too hard to sing. When "Over the Rainbow" won the Academy award in 1939, it was yet another signal that a flood of sentimentality—here gushing from Kansas rather than Oklahoma—was about to engulf New York urbanity, from top hat to tails.

Although he had other successful film songs, such as "Happiness Is a Thing Called Joe" (1943), Harburg found himself increasingly disgusted with Hollywood, not only for its commercialism but for the unwritten censorship that outlawed any social commentary. He found himself branded a "Red" because of the "Dime" song and because of his continuing support for Roosevelt and the New Deal which, as he himself put it, "scared the moguls to death." With foreshadowings of McCarthyism looming in Hollywood in the 1940s, Harburg returned to the Broadway theater. There his long-standing interest in letting songs unfold from character enabled him to adjust, as Howard Dietz could not, to the new "integrated" musicals. For nearly twenty years he wrote shows that dealt with issues of social significance, from women's liberation in *Bloomer Girl* (1944) to the antiwar protest of *The Happiest Girl in the World* (1961).

While successful as musicals, almost none of these shows produced a hit song—further testimony of the divorce between Broadway and Tin Pan Alley occasioned by the "integrated" musical. The exception to that testimony, of course, was *Finian's Rainbow* (1947), where Harburg collaborated with composer Burton Lane and also helped write the book. Yet even in this great show, the songs that became popular were not the acerbic critiques like "When the Idle Poor Become the Idle Rich" but the naughty "When I'm Not Near the Girl I Love (I Love the Girl I'm Near)"; not Harburg's Gershwinesque rhyming of "adorish/l'amourish/vanquish/relinquish/resish/relish/swellish condish," but the nostalgic "How Are Things in Glocca Morra."

The biggest hit from the show was, in fact, its least "integrated" song. It had been written earlier, for a movie where it was never used, and wound up in *Finian's Rainbow* only through the graces of Harold Arlen. Harburg and Lane had invited Arlen over one evening to hear their songs for the show, and Arlen praised all of them—except a sentimental balled called "We Only Pass This Way One Time." Harburg suggested that Lane play the unused movie ballad, and Arlen immediately dubbed it "champion stuff." "That decided me," Harburg recalled; he "tossed out" "We Only Pass This Way One Time" and wrote a new lyric, "Old Devil Moon."[15]

Arlen's preference for the music that became "Old Devil Moon" probably reflects its similarity to his own "bluesy" compositions. Harburg, too, sensed in Lane's melody "more of a Negro feeling," and started his new lyric by "looking for an idea, something that had to do with witchcraft, something eerie, with overtones of voodoo." It was not the black-magic trappings that made the lyric successful so much as Harburg's adoption of a casually vernacular style. The central phrase, "that old devil moon," a distinct echo of Mercer and Arlen's "That Old Black Magic," had precisely the same combination of exotic image—"devil moon"— and the nonchalantly familiar "that old. . . ." Such a contrast of ordinary and extraordinary goes all the way back to Porter's laconic plea, "do do that voodoo that you do so well," which spawned a coven of necromantic lyrics, from "Bewitched" to "Witchcraft." As the lyric unfolds, Harburg's language acquires a slangy elegance:

> You and your glance
> make this romance
> too hot to handle.
> Stars in the night
> blazing their light
> can't hold a candle
> to your razzle dazzle.

Here he weaves together heat and light imagery and subtly plays true rhymes against off-rhymes—"glance/romance/handle/candle" and "blazing/razzle/dazzle."

Most intricate of all is the way Harburg's colloquial phrasing matches the driving music right down to the "razzle dazzle" that follows the music's abrupt descent back to the final A section.

Just as that musical pirouette ties the end of the release to the beginning of that final section, Harburg uses "you" ambiguously, first seeming to complete "to your razzle dazzle, you!" but then sliding into a contraction to start the final phrase, ". . . you——'ve got me flyin' high and wide." When Lane's music takes still another unexpected turn just before the close, Harburg breaks out in a rare burst of slang: "wanna cry, wanna croon, wanna laugh like a loon." Harburg's use of such driving repetitions to express the delightful torments of romantic possession matches Lane's insistent—but never monotonous—musical repetitions. Given an unusual musical structure—no verse and a forty-eight-bar chorus—and what Alec Wilder praises as Lane's "dangerous but in this case successful use of the mixolydian mode"[16]— Harburg may have abandoned his usual attempt to ground the lyric dramatically. Whatever the reason, "Old Devil Moon" transported Harburg from his normally elevated poetic style down to an enchanting earthiness. While he wrote successfully for the theater for years afterwards, "Old Devil Moon" was Harburg's swan song as a popular lyricist, an "unpragmatically" pure Tin Pan Alley song, devoid of any but the slimmest sense of character or setting, and thus easily detachable from one of the most tightly integrated musicals.

Fine Romances: Dorothy Fields and Leo Robin

I'm not out to write popular hits, though I've written songs that have become popular.

DOROTHY FIELDS

Richard Rodgers ruefully remembered how, on his last day in Hollywood, he called on his producer, Irving Thalberg, to say goodbye. After a few minutes of awkward conversation, Rodgers suddenly realized that Thalberg did not know who he was. The incident was the last of a series of bitter frustrations for Rodgers in Hollywood, and it typified what he regarded as the anonymity of an assembly-line industry. The songwriting teams that did well in films, he concluded, were usually ones that had not come from Broadway, where composer and lyricist work closely with cast, director, and choreographer right down to opening night.

While Rodgers' conclusion certainly applies to many theater writers, most notably Oscar Hammerstein, it overlooks the fact that the Gershwins wrote some of their best popular songs in Hollywood. Cole Porter, too, had success in films, though even by his own admission, it was by writing down to the lowest common denominator of Hollywood standards. A more significant exception to Rodgers' observation could be made for two lyricists,

Dorothy Fields and Leo Robin, who brought some of the witty urbanity of the golden age to film songs during the 1930s. The exception would need to be qualified, however, since while both lyricists began writing in New York in the late 1920s, their theatrical experience had not been very extensive by the time they pulled up stakes for the West Coast.

According to Fields herself, her maiden effort at lyric writing, "Diga Diga Doo" (1928), nearly ended in disaster. Written for a revue at the Cotton Club, the song was sung, "coon" song fashion, by Adelaide Hall while a bevy of black chorus girls danced in two-piece red-sequined Zulu costumes. On opening night, Fields recalled, her father, the great vaudevillian Lew Fields, was present, along with "my family, my friends, the press, and, all-powerful at the time, Walter Winchell." When Hall sang "Diga Diga Doo," Fields claimed, she substituted "way-out, suggestive, risqué lyrics of someone else's devising, and my father practically froze in his seat."

> He demanded to know whether I wrote those words, and of course I told him I hadn't. At the end of her number, my father, the great Lew Fields, took the floor to applause, and made the announcement that I—his daughter—had *not* written those particular lyrics.[1]

One wonders whether the self-styled "lady lyricist" doth protest too much in this recollection. Adelaide Hall did not have too push Fields' lyric very far, since the title phrase, couched in such lines as "when you love it's natural to—Diga Diga Doo" and "how can there be a Virgin Isle with Diga Diga Doo?" is a typical '20s euphemism for sex (like "makin' whoopee"). What's more, in the lyrics she wrote for other Cotton Club shows and for Hollywood movies of the 1930s, Fields displays a sexy forthrightness, the winking naughtiness of "Diga Diga Doo" evolving into a slangy sensuousness all the more erotic for being so casually put.

Such passionate nonchalance may reflect Fields' stance as a female lyricist, the calculated "curve" she gave to saying "I love you" in thirty-two bars. She was clearly aware of her anomalous position as a woman on Tin Pan Alley. Though she always cited others, like Dorothy Donnelly and Carrie Jacobs-Bond, such women wrote in the traditional, elevated style. Donnelly supplied the soaring sentiments for such Sigmund Romberg operettas as *Student Prince,* and Bond's big hits were the maudlin "I Love You Truly" and "Just A-Wearyin' for You."

Fields, on the other hand, opted for the colloquial, urbane style of Hart, Gershwin, and Porter, always giving it her coolly sensuous twist. In a profession that required shoulder-to-shoulder collaboration with cigar-smoking men, most of whom thought a woman's place was in front of the footlights, Dorothy Fields was equally adept at turning a simple slang catch-phrase to passionate ends or spinning a list of metaphors laced with flippant rhymes. She did so, moreover, in collaborations with numerous composers with widely varied musical styles. So adept was she at holding her insouciant ground that she could even make Jerome Kern swing.

Although she had the advantage of coming from a theatrical family, her parents discouraged her from show business. She started out, properly enough, as a teacher at the Benjamin School for Girls in New York, and there she took part in amateur musicals written by her brother Herbert and the fledgling team of Rodgers and Hart. Like Ira Gershwin, Yip Harburg, and Howard Dietz, she also wrote "smarty verses" for magazines and managed to place poems in F. P. A.'s "Conning Tower." On the strength of that literary success, composer J. Fred Coots suggested she try writing song lyrics. Initially she found it a lot more difficult than poetry, admitting that the first lyric she "pieced out" to one of his melodies was "terrible."[2] Not dissuaded, Coots got her a job in Tin Pan Alley, at the Mills publishing house, where she blew her first assignment—writing a lyric for a song about aviator Ruth Elder, who was planning to fly across the Atlantic. "Ruth Elder never made it," Fields recalled, "and neither did my song." In 1927, however, she met another composer at Mills, Jimmy McHugh. McHugh had started out in Irving Berlin's Boston office as a "bicycle-plugger," pedaling around town with a keyboard on his handlebars, singing songs in the Berlin catalog. When he moved to New York, he began writing his own music and had hits such as "When My Sugar Walks Down the Street" (1924) and "The Lonesomest Girl in Town" (1925). He also wrote music for revues at the Cotton Club, which by 1927 was featuring Duke Ellington, and McHugh siphoned off some of the rangy, driving flavor of Harlem jazz for his own songs.

In Dorothy Fields, McHugh found someone who could create a sensuous, slangy lyric style perfectly suited to such music. She and McHugh adapted their Cotton Club shows for Broadway as a series of *Blackbirds* revues, and it was in the *Blackbirds of 1928* that the pair had their first big hit, "I Can't Give You Anything But Love." Actually the song was written back in 1926, but the story of its *Blackbirds* genesis is part of Alley lore:

> They racked their brains for days for an idea but drew nothing but blanks. Then one evening, while walking down Fifth Avenue, they noticed a young couple window-shopping in front of Tiffany's. It was obvious they didn't belong to the carriage trade . . . as the songwriters drew nearer, they heard the young man say: "Gee, honey, I'd like to get you a sparkler like dat, but right now, I can't give you nothin' but love!" Then and there the team of Fields and McHugh broke all speed in getting to a Steinway, and inside of an hour, they completed the smash song for which they had been searching.[3]

So, as they say, the story goes. However they got the wonderful catch-phrase title, the verse looks suspiciously like a Rodgers and Hart song of 1926, "Where's That Rainbow?":

> My luck will vary surely,
> that's purely a curse.
> My luck has changed—it's gotten
> from rotten to worse.

While Hart's phrasing here is stilted, Fields renders the same idea in thoroughly conversational cadences:

> Gee, but it's tough to be broke, kid.
> It's not a joke, kid—it's a curse.
> My luck is changing—it's gotten
> from simply rotten to something worse.

She handles slang so deftly one can almost hear a Brooklynese undercurrent in "It's a coise!"

Not only is her phrasing utterly natural, Fields creates a Damon Runyon character to voice it. His diamond-in-the rough poignancy shines out in the chorus, where he switches from "kid" to "baby," placing the harshly tender tag over two kicking notes. By rhyming "I can't give you anything but *love,* baby" with "That's the only thing I've plenty *of* baby," then accenting the rhyme with high notes, Fields reflects the singer's brassy pauperdom with a flourish that shows an equal paucity of rhyme. She deepens the wry portrait of this impoverished Romeo by limiting his imagination: instead of longing for jewelry *from* Cartier's or Tiffany's, he only specifies where they must *not* originate:

> Gee, I'd like to see you looking swell, baby,
> diamond bracelets Woolworth doesn't sell, baby

By suppressing syntactic connections, Fields deepens the conversational feel and creates a character who even economizes through telegraphic understatement of his feelings.

Before leaving Broadway for Hollywood, Fields and McHugh wrote three more revues, *Hello Daddy* (1928), *Vanderbilt Revue* (1930), and *International Revue* (1930). In the last of these, they had their biggest hit, "On the Sunny Side of the Street," where Fields once again demonstrated her deft handling of slang phrasing. Her abrupt imperatives punctuate McHugh's "bumptious" melody and highlight its six interval leaps with lyrical reaches: "Grab your coat, and get your HAT!" and "Can't you hear a pitter-PAT!" In the release, however, she uses that musical springboard to launch a jaunty boast—"If I never have a CENT I'll be rich as Rockefeller"—a boast made even more brash by her telegraphic omission of syntactic connections:

—gold dust at my feet
on the sunny side of the street

Such slangy leaps elicit praise from Alec Wilder: "Her lyrics often swung, and their deceptive ease gave a special luster to McHugh's music."[4]

It was in Hollywood that Fields turned her skills with slang phrasing to more sensuous ends, both in her collaborations with McHugh and with Jerome Kern. "Don't Blame Me" (1932), for the film *Dinner at Eight,* marks the shift to a more languorously erotic style. Fields makes the prosaic catch-phrase, "don't blame me," a smoldering match for McHugh's three opening half-notes, each word offering a different open vowel—*o, a, e,*—to provide a perfect vehicle for the torchiest of singers. But lest that singer get too ardent, Fields keeps the mood insouciant by using McHugh's triplets that break up potentially sentimental phrases like "for *fal-ling-in* love with you" and "I'm *un-der-your* spell." In this lyric, too, we find Fields experimenting with subtle rhymes that keep her phrasing wryly off balance:

I'm under your sp*ell, but*
how can I h*elp it?*

As usual, however, it is the vernacular ease—here with an exasperated touch—that takes precedence over clever rhyme.

Similarly sensuous nonchalance marks "I'm in the Mood for Love," from the 1935 movie *Every Night at Eight.* Here Fields lifts another ordinary catch-phrase, this time one from the prosaic realm of menu selection, and turns it to passionate ends. To keep the mood simmering—but not explosive—she douses the invitation with slangy interjections—*"simply* because you're near me, *funny,* but when you're near me" and *"Oh!* is it any wonder I'm in the mood for love?" Such matter-of-fact passion, all the more erotic for its casual understatement, reaches its climax with a conversational afterthought:

If it should rain we'll let it
but for tonight, forget it!

Paradoxically, Fields evokes far more sensuality with such off-handed slang phrases than a lyricist like Howard Dietz does when he reaches for the tempestuous heights.

Dorothy Fields could even sound sensuous when she wasn't writing love songs. In "I Feel a Song Coming On," for example, she lifts a catch-phrase from the most unromantic context of incipient illness and transposes it to one of joy so infectious it seems erotic. She intensifies McHugh's syncopated melody by collapsing the last two syllables of "victor-*ious*" and "glor-*ious*" into a contagious "yas," which then spreads to "warning ya." In the release, the same sounds multiply and divide in different combinations, as "story" and "glory" echo the middle syllables of "victorious" and "glorious," while "ringin' thru ya" and "Hallelu-jah" go back to "ya." As such infectious sound play takes over the lyric, it exemplifies the implied metaphoric connection between spreading passion and general contagion.

In 1935 Jerome Kern came to Hollywood to adapt his Broadway hit *Roberta*, and McHugh and Fields, as established film song-writers, were called upon to help him with the transition from stage to screen. One of Dorothy Fields' first assignments was to write a lyric for a new Kern melody, "Lovely to Look At," a tricky task since the music consisted simply of a chorus that was only sixteen bars long (instead of the usual thirty-two). She got a little more lyrical elbow room when Fred Astaire, the film's star, asked Kern for a fast-paced number, illustrating his request (so the story goes) by tapping out the rhythm he wanted with his feet. Kern accommodated him by revising "I Won't Dance," a song that he had written with Otto Harbach and Oscar Hammerstein for a 1933 English show, *Three Sisters*. The rhythmic revisions, however, necessitated some lyrical ones. Fields revamped the original, keeping the punchy opening line—"I won't dance—don't ask me"—but from then on she gave it her own urbanely brassy eroticism:

> when you dance you're charming and you're gentle!
> 'Specially when you do the "Continental."

> But this feeling isn't purely mental;
> for heaven rest us,
> I'm not asbestos

Fields's skill in this passage is particularly deceptive since the music is, as Gerald Mast notes, "bizarrely meandering," the "most difficult release Kern ever wrote."[5] The lyric, nevertheless, sounds perfectly off-handed, even chatty. The final twist she gives to the terpsichorean refusal turns apparent shyness into uncontrollable passion: "so if I hold you in my arms . . . I won't dance!"

Working alone with Kern on another Astaire-Rogers film, *Swing Time* (1936), Dorothy Fields took the same clash of frigidity and passion and turned it into a brilliant list song whose catalog of images and allusions rivals Porter at his tops. Subtitled "A Sarcastic Love Song," "A Fine Romance" inventories not a list of compliments but one of complaints that border on insult. The woman's half of the duet is particularly deft since it reverses the traditional male complaint about female frigidity:

> we should be like a couple of hot tomatoes,
> but you're as cold as yesterday's mashed potatoes

After raising the specter of her own spinsterhood—"I might as well play bridge with my old-maid aunts"—she picks up on the cold food motif by finding "Jello" preferable to romance with her "good fellow," then turns to an image that combines emotional cold with motionless calm:

> you're calmer than the seals in the Arctic Ocean,
> at least they flap their fins to express emotion

She caps the list with a risqué image of unflappable cool: "I've never mussed the crease in your blue serge pants."

Along with clever images, Fields manages her patented effects of stylish slang:

> a fine romance! with no quarrels,
> with no insults, and all morals!

The title phrase itself is a clever reworking of the slang formula for "a fine kettle of fish" with a playful turn on "fine" to suggest the genteel lover whose refinement she finds so exasperating. Using Kern's music to accent that frustration, Fields rags the very word "romance" from r*omance* to *ro*mance. That distortion of accent creates a back-to-back pattern of iambs and trochees:

> a *fine*/ *ro*-mance!
> with *no*/ *kiss*-es!
> a *fine*/ *ro*-mance,
> my *friend*/ *this* is!

The rhythmic pattern, in turn, enforces the sharp contrasts of imagery, as does the musical device of moving the melody—and the singer's sensuous exasperation—higher and higher with each pair of notes.

While the witty catalog of images foregrounds itself in "A Fine Romance," in "The Way You Look Tonight," another hit from the film, Fields uses an utterly simple lyric to bring out the elegant sensuousness of Kern's gracefully insistent melody. The title phrase signals her nonchalantly passionate angle—not the sumptuous praise of eternal devotion to immutable beauty but the off-handed, snapshot compliment to *just* (meaning both "merely" and "precisely") "the way you look tonight":

> with your smile so *warm*
> and your cheek so soft,
> there is nothing *for m-*
> e but to love you

The barely heard tie-rhyme (*warm*/*for me*), the chattily resigned "there is nothing for me but to . . . ," and the lightly sensuous "so warm . . . so soft" all underscore the easy grace Astaire brought to a song. Even when she borders on the histrionic, with a line like "never, never change," Fields sure-handedly follows up with the casual slang of "Won't you please arrange it?" In yet another song from *Swing Time*, "Never Gonna Dance," a reworking of "I Won't Dance," Fields combined the colloquial with the romantic in a phrase that perfectly characterizes Astaire and Rogers: "la belle, la perfectly swell romance."

Although she continued to write for Hollywood, Fields began gravitating back toward the Broadway theater in the late 1930s, a move that saw her production of popular songs—and for a time any kind of songs—rapidly diminish. Teamed with her brother Herbert, she wrote the book—but not the lyrics—for such Cole Porter musicals as *Let's Face It* (1941) and *Something for the Boys* (1943). Porter himself, always the lyricist rather than the librettist, asked in amazement, "You Fieldses want to write *book*?" Asked if she found it difficult "to retire as a lyricist and bequeath that spot to Porter, Dorothy Fields replied, "Oh, honey, let me tell you, it's great . . . the book is always the toughest thing to do." In 1946, Dorothy and Herbert Fields had their greatest "book" triumph in *Annie Get Your Gun,* with Irving Berlin providing the songs.

Dorothy Fields went on to write both "book and lyrics" for such musicals as *A Tree Grows in Brooklyn* (1951) and *By the Beautiful Sea* (1954). While the shows were successful, however, the meditative, philosophical songs, such as "Growing Pains" and "I'll Buy You a Star" never established themselves as independent hits. By this time, however, Fields had come to regard herself purely as a Broadway "book-writer." Rather than trying to write "popular song hits," she claimed, "I'm writing a song to fit a spot in the show. To fit a character, to express something about him or her . . . to move that story line forward."[6]

Yet a few of those thoroughly "integrated" songs did become popular—at a time when it was a rarity for a Broadway musical, even the most successful, to produce an independent hit. When Dorothy Fields collaborated with composer Cy Coleman on *Sweet Charity* in 1966, for example, the show produced not one but two such hits—perhaps in part because Fields reverted to her early Cotton Club style. *Sweet Charity*'s low-life setting, its dynamic central character, and Coleman's brassy music called forth Fields' skill with slang, from the mixture of high and low diction in "(Hey) Big Spender,"

The minute you walked in the joint,
I could see you were a man of distinction,
a real big spender,
good looking, so refined,

to the even more bumptious phrases of "If My Friends Could See Me Now":

I'd like those stumble-bums
to see for a fact
the kind of top-drawer first-rate
chums I attract!

The staccato exclamations, punctuated by rhymes like st*u*mble/b*u*ms/ch*u*ms, register Charity's crowing enthusiasm in her new-found state. It is easy to see Dorothy Fields herself in that character—starting out with earthy Harlem nightclub songs, moving up to urbanely enticing Hollywood lyrics, finally reaching the aristocratic heights of her profession as a thoroughgoing lyricist-librettist for some of Broadway's most integrated musicals, but still able, on occasion, to strut her slangy, sensuous stuff as a "pop" songwriter.

Like Dorothy Fields, Leo Robin broke into songwriting in New York in the late 1920s and then moved to Hollywood, where he brought a flair for casual sophistication to film songs. Starting out as a playwright, Robin looked up fellow-Pittsburgher George S. Kaufman when he arrived in New York and presented some of his manuscripts. "Kaufman," according to songwriter Edward Eliscu, "wasn't too impressed, but he was gentle about it":

He asked, "What else have you written?" Leo said, "Well, I've written some songs," but he was rather shame-faced about it. "Let's see 'em," insisted Kaufman, and Leo showed him one, a lyric he'd written called "My Cutie's Due at Two to Two Today!" and Kaufman immediately said, "That's it—*that's* what you should be doing!"[7]

Kaufman's advice quickly paid off. In 1927 Robin was paired
with co-lyricist Clifford Grey and assigned to put words to a
rhythmic melody Vincent Youmans had written years before as a
Navy song. Although they came up with yet another stiff and
stilted Alley version of a "spiritual," replete with such clichés as

> Satan lies awaitin'

and

> shoo the blues away,

"Hallelujah!" was a big hit in *Hit the Deck* and established Robin
as a lyricist.

On the strength of that success he signed a Hollywood contract with Paramount in 1929, but in his early songs Robin had to
suppress his lyrical flair to meet the banal studio standards.
Teamed with composer Richard Whiting on several Maurice Chevalier films, he wrote songs such as "Louise" (1929), which had to
have lines tailored to Chevalier's patented hand gestures: "Every
little breeze" for waving and for fluttering:

> Birds in the trees
> seem to twitter "Louise"

Occasionally, in a song like "My Ideal" (1930), Robin could reveal
flashes of his flippant wit, such as

> The idol of my heart
> can't be ordered à la carte.

For a non-Chevalier film, *Dance of Life,* Robin and co-lyricist
Sam Coslow turned out a very different sort of lyric for Whiting's
brassily bluesy "True Blue Lou":

> He gave her nothing, she gave him all.
> But when he had his back to the wall,
> who fought to save him,
> smiled and forgave him?

"The Blue Lou," however, while a jazzy relief from Chevalierese, was hardly in Robin's urbanely witty groove.

Nor was "Beyond the Blue Horizon" (1930), which he and Whiting wrote for *Monte Carlo,* a picture directed by Ernst Lubitsch. Lubitsch was one of a handful of movie men in the early days of film who experimented with musicals where the songs fit character and situation. (Others included Rouben Mamoulian, who later directed *Oklahoma!,* and Adolph Zukor, who at once impressed Robin by using the term "lyrics" rather than Hollywood's standard term, "the words.") Instructing Robin to write for "characters" rather than "performers," Lubitsch asked for a special song for *Monte Carlo*—an internal monologue for runaway princess Jeanette MacDonald to sing as she gazed out of a moving train window at peasants working in a field. Robin and Whiting came up with "Beyond the Blue Horizon," and Lubitsch used it for what Gerald Mast calls the "first sensational Big Number in an American filmusical, the first extended musical sequence in which song, story, style, and meaning combine with a pyrotechnical display of cinema's visual devices."[8] Yet precisely because of that "integration"—the music echoing the clicking train wheels, the camera cutting deftly from engine to wheels to window to field—Robin knew his lyric could not be obtrusively clever: "it *had* to be simple . . . not too sophisticated."[9] Instead of witty imagery and clever rhymes, which might upstage the cinematics, he relied on long *i* and *e* vowels and *n* and *z* consonant sounds to echo the expanding exultation of the princess, which unfolds with the train's forward, repetitive movement:

> My life has *only begun,*
> *beyo*nd the blue hori*zon*
> *lies* a r*ising* s*un.*

Gradually Hollywood gave Robin more chances to flash his wit, though such lyrics quickly earned him a reputation for being "arty." Teamed with Lewis E. Gensler in 1934, he hit his urbane stride with "Love Is Just Around the Corner," which Bing Crosby sang in *Here Is My Heart.* The title phrase gave a romantic twist to

Herbert Hoover's by-then laughable prediction of prosperity, the rhymes on "corner"—"sentimental mourner" and "couldn't be forlorner"—were refreshingly antiromantic, and there was a suggestive turn to "around" in the final invitation to "go cuddle in a corner" where "I'm around you." In the release, Robin turned one of the wittiest metaphors of the golden age:

> But strictly between us,
> you're cuter than Venus
> and what's more you got arms!

The elegant allusion, the archly casual "strictly between us," and the thumping slang of "what's more you got arms" gave Hollywood another taste of the blend of slang and sophistication that spiced the lyrics of Dorothy Fields.

Leo Robin found his perfect collaborator in 1934, when he teamed with Ralph Rainger, but at first the studios balked at their songs. "Love in Bloom" (1934), which Jack Benny later adopted as his theme song, was initially found "too high-class" by the studio bosses, and Alley publishers asked, "Who the hell can sing this? Too rangy musically, and the lyrics are too fancy."[10] Bing Crosby's recording made the song a hit, however, as did his version of "June in January" (1934). What probably alarmed Hollywood and Tin Pan Alley in these lyrics was Robin's use of metaphor, though a line like

> the snow is just white blossoms
> that fall from above

seems hardly too complex or daring for a song lyric. It does, however, resemble Imagist poetry with its delicate metaphors linking emotion to images in nature (blossoms, in fact, were a staple of Imagist poems). Robin's remark that he wouldn't be surprised to come across his title-phrase for "June in January" "looking through some old poetry written by some Chinese or Greek three thousand years ago" suggests he may even have been aware of Imagist adaptations of Greek and Chinese poems, as well as their translations and imitations of Japanese haiku. A famous haiku, cited by Ezra Pound for its subtle juxtaposition of images, is about the same seasonal reversal Robin elaborates in "June in January": instead of regarding the snows of winter as springtime foliage, however, the haiku poet cleverly "mistakes" a late-summer butterfly for one of those blossoms.

Robin created far more intricate imagery for "Thanks for the Memory" (1937), a list song that rivals any of Porter's as the epitome of nonchalant sophistication. In this lyric, the studios called for precisely the kind of lyric Robin wrote best. Mitchell Leisen, the director of *Big Broadcast of 1938,* asked Robin to write a song to solve a problem six scriptwriters couldn't cure with dialogue.

> Leisen says, "This is a young, sophisticated couple who had been married. And they meet by chance on an ocean liner. I want them to show they are still in love, but they dare not say it. It's got to be implied. Now if you guys can write a song like that, fine."[11]

Although he grumbled that "It's not easy to say 'I love you' without saying it," Robin set to work, but he had hardly begun before Leisen popped by to add another complication: "This song is going to be sung by Bob Hope. And while it's a serious song, a guy like that has to get laughs."

What the director had posed for Robin was precisely the same problem that Hart, Porter, and Gershwin had resolved for years—saying 'I love you' urbanely, wittily, and, somehow, without actually *saying* it. Robin clearly took Porter's list songs as his inspiration, mixing elegant European images with prosaic American ones:

> candlelight and wine,
> castles on the Rhine,
> the Parthenon
> and moments on
> the Hudson River Line

Working with the rising musical repetitions, Robin uses the final prosaic image to deflate rather than top his romantic series. In the second A section, similarly he builds suspensefully from "motor trips" to "burning lips," then cuts to "and burning toast and prunes."

The character Robin creates is an intricate transformation of the insouciant, broken-hearted lover of Porter's "Just One of Those Things"—an urbane sophisticate who feels compelled to deprecate a sentimentalilty he is equally compelled to indulge. In the release, appropriately, Robin allows such a figure to come closer to direct emotional expression, yet he still keeps up his mordant guard:

> we said good bye with a highball;
> then I got as "high" as a steeple.
> But we were intelligent people;
> no tears, no fuss,
> hurray for us

Ironically, it is only with alcohol that sober grief is released, only to be repressed immediately by sophistication.

In the final A section, Robin moves this desperately elegant sensibility as close as it can come to emotional expression:

> so thanks for the memory,
> and strictly entre nous,
> darling, how are you?
> And how are all the little dreams
> that never did come true?

From the artificial intimacy of "entre nous," slangily framed by "strictly" (a reworking of his own "strictly between us you're cuter than Venus"), Robin shifts to "darling" as at once the most openly affectionate yet affectedly sophisticated of addresses. The "How are you" that follows thus is transformed from the utterly banal to the guardedly tender, hardly the thing that needs to be whispered "entre nous" in its normal context of everyday greeting but here, so framed, the tip of the emotional iceberg.

"Thanks for the Memory," to the utter amazement of Robin and Rainger, brought tears to the eyes of all the studio bosses and went on to win the Academy award as best song. Despite their sophistication, however, the songwriters were still perfectly capable of turning out pure Hollywood schlock, such as "Blue Hawaii" (1937), which the composer himself lambasted as a "piece of crap," even though it turned out to be the biggest of all their hits. Their successful collaboration ended in the early 1940s, when Rainger moved from Paramount to Twentieth Century-Fox, though Robin continued writing with various composers, deftly adapting his lyrical style to their music. With Jerome Kern, for example, he had a sentimental success, "In Love in Vain" (1946), then in 1948 he matched Harold Arlen's swinging sophistication with the exuberant ennui of

> Hooray for love—
> Who is ever too blasé for love?

Like Dorothy Fields, Robin showed that he could make the transition from film songs to integrated musicals, but his return to the stage in 1949 was a brief one. Even then he had to be goaded into it by his neighbor, composer Jule Styne, who knocked on Robin's door one day and proposed collaboration. Styne was willing to leave a successful career as a film composer in order to take on the challenge of Broadway, but Robin was far more reluctant. "He was making quite enough money writing film songs," Styne recalls, "And his two previous attempts at writing for the stage hadn't been successful, the last one [*Nice Going*] a typical Boston massacre."[12]

What appealed to Robin, however, was that Styne wanted to transform Anita Loos' play, *Gentlemen Prefer Blondes,* into a musical. The character of Lorelei Lee—"terribly sexy, exotic, a cheap kept-lady"—intrigued Robin, since it afforded him a chance to mix brassy slang and sophisticated wit, much as Dorothy Fields did in *Sweet Charity.* One of those songs, even though rooted in that character, became an independent hit:

> men grow cold, as girls grow old
> and we all lose our charms in the end,

but square-cut or pear-shape,
these rocks don't lose their shape

"Diamonds Are a Girl's Best Friend" made a star out of Carol Channing as a daft but not-so-dumb blonde, delivering such lines with a bump and grind to emphasize the pun on "end."

With *Gentlemen Prefer Blondes,* Robin and Styne established themselves on Broadway, but ended their collaboration. Styne went on to compose for such successful musicals as *Gypsy* and *Funny Girl,* one of the last "pop songwriters," as Martin Gottfried notes, to cross "show business's great divide—from Tin Pan Alley to Shubert Alley, from pop music to theater music."[13] Robin, however, returned to Hollywood. What little songwriting he did in the years that followed was primarily for films, where he was content to turn an occasional phrase, such as "A Little More of Your Amor," that echoed, if ever so faintly, the wit he brought to film songs in the golden age.

Hip, Hooray, and Ballyhoo: Hollywood Lyricists

> *Out here in Hollywood a songwriter was always the lowest form of animal life.*
>
> HARRY WARREN

Among the many New York writers imported by Hollywood in the wake of the first "talkies" was Dorothy Parker, and one of her assignments was to supply lyrics for movie theme songs. At first she thought the studios wanted the same wit she flashed in her society verse, but she quickly realized she was in under her head. What Hollywood demanded was not deft rhymes and clever imagery but lyrics simpler than the most banal Alley fare. Told to write a lyric for the theme song to a movie called *Dynamite*, Parker sarcastically turned in "Dynamite, I Love You"—then packed her bags and headed back east.

Parker's lyric reduced the Hollywood formula for songs to its absurd minimum—a formula, according to Ethan Mordden, that was "pop music at its heart, so universal in appeal" that it lacked "the penetration of individuality . . . the peculiar individual edges that produced, for example, Porter's tortured, Hart's puzzled, Berlin's incurious, or Ira Gershwin's chaste love lyrics."[1] What Hollywood wanted instead was a pure Tin Pan

Alley formula, watered down and transplanted on the West Coast.

The link between Tin Pan Alley and Hollywood went all the way back to the earliest days of film. As nickelodeons—movie theaters that charged a nickel—spread across the country in the first decade of the twentieth century, Tin Pan Alley's "song pluggers knew they had acquired a new shop in which to exhibit their wares":

> They did so with song slides, song fests, and amateur nights. Song pluggers often invaded the auditoriums themselves to provide live entertainment by singing the current numbers from their respective firms. By the early 1910s, one plugger, Sammy Smither, boasted he could plug a song fifty times an evening by making the rounds of nickelodeons.[2]

Soon pluggers were obtruding their products into the films themselves, bribing the piano player who accompanied silent films to use his firm's songs as "background music." Among his many contributions to the development of the art of film, D. W. Griffith tried to put a stop to this practice: *The Birth of a Nation* was the first film to have its own background music specifically composed for the picture.

Undaunted, the publishers turned to other ways of cashing in on films. Just as Tin Pan Alley had impeded the development of the "integrated" stage musical, it kept Hollywood from evolving a distinctive song style appropriate to the new medium of film. During the golden age, most movie musicals, like most Broadway shows, consisted of a string of standards that said "I love you" in thirty-two bars. Even before the advent of sound pictures, Tin Pan Alley publishers and Hollywood studios had invented the "theme" song—a song with the title of the movie, whose melody was played over and over as background music. Such theme songs paid a double bonus. Tin Pan Alley publishers found that a successful movie could boost a song's sheet music sales; Hollywood producers found that a successful title song could help promote a movie. Even when the song title and movie title were not the same, as with the song "Charmaine" for the 1926 silent film *What Price Glory,* the linkage reaped profits on both coasts. Nothing illustrates the marriage of Tin Pan Alley and Hollywood more clearly than the promotional broadcast for "Ramona," the most successful movie theme song of the 1920s: it was introduced over coast-to-coast radio with the soloist, Delores Del Rio, singing from Hollywood, and the accompaniment, played by Paul Whiteman's orchestra, emanating from New York.

Thus when sound did come to the screen, with *The Jazz Singer* (1927), the first successful film to employ singing, and its even more influential follow-up *The Singing Fool* (1928), Tin Pan Alley and Hollywood were firmly wed. Because film songs were based upon the simplest of Alley formulas, Hollywood quickly became the major outlet for Tin Pan Alley wares—new and old. The inclusion of established hits—Jolson singing "Blue Skies" or Fanny Brice singing "Second Hand Rose"—could boost a picture at the box office. Shrewdly realizing that fact, Warner Brothers, a minor studio with Rin Tin Tin as its only bona fide star until it broke into sound, quickly bought up the publishing firms of Witmark and Sons, T. B. Harms, Remick's, and the full catalog of DeSylva, Brown, and Henderson songs.

Other studios followed suit, and in a few years almost all of the old Alley houses were owned by Hollywood. By the 1930s the child had devoured the parent; the old publishers on Tin Pan Alley complained that the Hollywood studios now controlled the production of new songs. The role of the publisher as one who sought out new composers, supervised the assembly-line production of songs, and then disseminated them through a network of pluggers had come to an end. "All we do," one publisher lamented, "is collect royalties."

If anything, Hollywood looked upon music even more commercially than the most ruthless Alley publisher. Not only did it cram its newly acquired backlog of hits into movies, it transported Tin Pan Alley's wordsmiths and tunesmiths from their cramped quarters along 28th Street and plopped them down in spacious studios—sometimes with windows that gave a view of the mountains ("just the atmosphere for the composition of melodies,"[3] cooed one executive). If Dorothy Parker found the situation laughable, it is easy to see why songwriters like Rodgers and Hart were so frustrated in Hollywood. They had hoped films could be used to develop a new kind of musical, more informal and casual, where songs not only grew out of character and situation but were woven together by rhymed and rhythmic dialogue. What the studios wanted, however, was simply a string of hits, performed with no relation whatsoever to dramatic context.

Hollywood's reluctance to use songs more integrally was not, however, purely commercial. It stemmed, in part, from the awkwardness of having characters suddenly burst into—and out of—song on the screen, without even the applause that cushions such outbursts in stage musicals. While Broadway moved, ever so slowly, toward the "integrated" musical, "never, *never* did Hollywood get over its horror of characters singing for purposes of plot."[4] When an actor sang in an early film, it was usually because he was playing the character of a singer—like Jolson in *The Jazz Singer*—and the song was performed as a "performance"—on a stage and before other actors playing the part of an audience. In Hollywood musicals, as Gerald Mast notes, "musical performers perform—and a nonmusical performer is an oxymoron."[5]

What such "performers" performed, naturally, were popular songs, so that, in a way, Tin Pan Alley's songs were perfectly appropriate for film musicals. The best way to showcase such "performances," moreover, was to make movie musicals about the making of Broadway musicals—"backstagers" as we now call them. The first full-scale film musical, MGM's *Broadway Melody of 1929*, set the formula:

> What was to become convention in every backstage musical is present here in pristine simplicity: the imperious impresario, his cohorts of yes-men, the dilettante backers, the star of the show breaking her leg at rehearsal, her replacement wowing the audience by apparently not even singing a note or swinging

a limb or doing anything else except simply stand there, and the big-hearted heroine surrendering her own chance of stardom and happiness for the sake of her kid sister.[6]

Rather than follow the usual practice of using established Tin Pan Alley hits, however, producer Irving Thalberg struck upon the innovation of having new songs written for *Broadway Melody*. He commissioned a young team, Arthur Freed and Nacio Herb Brown, to supply them. Although Freed and Brown were primarily West Coast Songwriters (Brown had never set foot in Tin Pan Alley), they nevertheless had all the old Alley formulas down pat: simple rhymes and harmonies, back-to-back box-like phrases, and sentiments that never ventured beyond the tritest romantic clichés. Even though they wrote most of their hits in the 1930s, Freed and Brown's songs *sound* as if they were written in the 1920s, at the height of the era of "bone-simple" music and lyrics. So perfectly anachronistic are such songs that they were later used in *Singin' in the Rain* (1952), a film musical (produced by Freed) about the birth of the talkies in the mid-1920s.

"You Were Meant for Me" (1929), for example, has the same rickey-tick structure as "Tea for Two," right down to the *ee/oo* rhymes and the simplest syntactic reversals: "I was meant for you." Those same features turn up again in "All I Do Is Dream of You" (1934) and, yet again, in "You Are My Lucky Star" (1935):

    you are
    my lucky star;
    I'm lucky
    in your arms

In 1936 Freed and Brown were at it again with the back-to-back repetitions of "(I Would) Would You." The very datedness of such lyrics give them their charm as period pieces from the heyday of Tin Pan Alley.

Such songs could be sprinkled across films with utter disregard for the dramatic context. On one of the rare occasions when Freed and Brown did write a song to fit a character and situation,

the results were so absurdly histrionic that one can see why Hollywood abhorred "integrated" songs:

> you came!
> I was alone!
> I should have known!
> You were temptation!

Even Bing Crosby himself laughed off "Temptation" (1933) and
its melodramatic setting in a Tijuana dive:

> "Temptation" was my first attempt at presenting a song dra-
> matically. . . . As a derelict I sang "Temptation" to a glass of
> tequila, while tears dripped into my beard from the circles
> under my eyes. Through trick photography a dame's face ap-
> peared in the glass of tequila; then, at the end of the song, I
> flung the glass at the wall and staggered out into the night. It
> was all very Russian Art Theater."[7]

Perhaps we should be grateful that Hollywood songwriters seldom gave in to the temptation of dramatic integration.

Freed and Brown were only one of numerous songwriting
teams who supplied Hollywood with songs in the old Alley style.
Almost as successful were Bert Kalmar and Harry Ruby, and
their biggest hit, "Three Little Words" (1930), typifies that style.
Ruby's melody, in fact, posed a seemingly insurmountable problem for a Hollywood song: its main phrase did not fit the usual
cookie-cutter pattern that always enabled a lyricist, as Dorothy
Parker did in "Dynamite," to say "I love you." Since it consisted
of a four-note phrase, it was too long for the three-syllable standard "I love you." What they needed, in other words, were *four*
little words, but they solved their problem—ingeniously, for
Hollywood—with the four syllable equation

> three lit-tle words
> which simply mean
> "I love you."

Given the plethora of such film songs, those that Berlin and the Gershwins supplied for Fred Astaire seem all the more impressive in their artful simplicity, as do the urbanely witty lyrics of Dorothy Fields and Leo Robin. An equally significant exception to the studied banality of most Hollywood lyrics were those Al Dubin and Mack Gordon set to the music of composer Harry Warren. In his survey of American popular music, Alec Wilder takes note of only one song by Freed and Brown and utterly ignores Kalmar and Ruby, yet he gives extensive treatment to Warren, hailing him as one of "the foremost pop song writers."[8] Like everyone, including Warren himself, Wilder laments the composer's near-anonymity; there is probably no greater indication of Hollywood's obliteration of individuality than the fact that a composer who wrote nearly as many familiar hits as George Gershwin is almost utterly unknown by the people who know his songs.

Warren had started out as an Alley composer, garnering his earliest hits with place-name pieces, from "Rose of the Rio Grande" (1922) to "Nagasaki" (1928). By the early 1930s, however, his songs such as "By the River Sainte Marie" and "I Found a Million Dollar Baby in a Five and Ten Cent Store" (1931), began to pick up what Wilder calls their "riffish" character—the same rhythmic, driving quality one finds in the music of Harold Arlen and George Gershwin. Such a style has a jazzy, New York flavor, and it is ironic that, while Harry Warren was Hollywood's most prolific composer, he remained a die-hard New Yorker. After his first stint on the West Coast in 1929, for example, he fled back to New York, explaining, "I missed Lindy's." In 1932, however, he returned to Hollywood and was teamed with lyricist Al Dubin—seemingly the least suitable collaborator for a composer whose melodies were becoming increasingly sophisticated. A seasoned Alley wordsmith, Dubin had done everything from immigrant nostalgia—" 'Twas Only an Irishman's Dream" (1916)—to sentimental ballads—"Just a Girl That Men Forget" (1923). His early film songs, set to Joseph Burke's typically rickey-tick tunes, vacillated between the dim-witted ebullience of "Tip Toe Through the

Tulips" (1926) to the maudlin stoicism of "Dancing With Tears in My Eyes" (1930). Thus he hardly seemed the lyricist to match Warren's driving melodies with energetic slang phrasing.

The pair was put to work at Warner Brothers by Darryl F. Zanuck on *Forty-Second Street*—another "backstager" that celebrates "hot-shot, ace-high, lowdown, dirty, crazy New York show biz," and Dubin and Warren came up with songs to match its "gritty vernacular ambience."[9] The film also marked the beginning of a five-year collaboration with choreographer Busby Berkeley at Warner Brothers. Because they were "Tin Pan Alley tunesmiths rather than Broadway composers," Gerald Mast argues, the songs of Dubin and Warren captured the "energy" of Berkeley's choreography. Warren's "eight-bar phrases build a thirty-two-bar refrain in sharp, bold outlines, a firm musical container for the liquid of Berkeley's ever dissolving visual figures."[10] Dubin's lyrics, one should add, were equally effective in keeping the numbers down to earth. "You're Getting to Be a Habit With Me," for example, gave a street-wise, urbane twist to Hollywood's formulaic "I love you." Dubin uses Warren's repetitive melody to portray a fated victim of romance, letting the numbingly insistent triplets underscore the metaphoric addiction:

> Ev'ry kiss, ev'ry hug
> seems to act just like a drug.

By the time he gets to the "now I couldn't do without my supply," the drug metaphor has paved the way for the brazen, street-talk proposal, "I must have you ev'ryday."

While such songs matched the New York atmosphere of *Forty-Second Street,* they were not "integrated" into plot and character but "show-stopper" production numbers; according to Harry Warren, they were completed even before a screenplay was written. The same free-standing character marked the many other songs Dubin and Warren turned out in the 1930s for other Warner Brothers "backstager" musicals. "I Only Have Eyes for You" (1934), for example, was interpolated into more than half a dozen different films. What lifts such a lyric above the usual run of film songs is Dubin's ability to match Warren's insistent melody with casually conversational phrases:

> Are the stars out tonight?
> I don't know if it's cloudy or bright,
> 'cause I only have eyes for you

Beneath such casually understated passion is an emotional progression intensified by insistent rhymes such as the *I/eye* in the title phrase and "*For you*" with "*or* on a crowded ave*nue*." He drives the lyric more forcefully still by following "ave*nue*" with "*You* are here, so am I," just as Warren's music pushes the end of the release into the final A section.

Even more slangily sophisticated is "Lullaby of Broadway" (1935), which earned Dubin and Warren their first Oscar. Once again, it was the idiom of New York that inspired Warren's "riffish" melody, and Dubin matched those rhythmic, repeated musical phrases with quick, cosmopolitan allusions—rhyming "taxis" with "Angelo's and Maxie's"—and bits of urbane argot—"the hip hooray and ballyhoo"—tossed off with laconic detachment. The voice of the song is a sophisticated, bemused observer of the New York scene, the perfect complement to the rhythm of the music which captures—yet keeps urbanely under control—the frenetic nightlife it describes. That voice also supports the witty transformation of the "lullaby" motif into "babes" and "sugar-daddies":

> "Hush-a-bye, I'll buy you this and that,"
> you hear a daddy saying,
> and baby goes home to her flat
> to sleep all day

Dubin skillfully uses the sharp musical accents to punctuate his vowels—"hush-a-*bye I'll buy* you"—but when Warren suddenly shifts to languorous "lullaby" notes Dubin draws them out: "*sleep tight ba-by.*" The brief lullaby interlude is then clipped off with a rhythmic flourish, which Dubin fits with the hard-edged alliteration and short vowels of "milkman's (instead of the more euphonious "sandman") on his way."

Even a song that could easily have become a sentimental paean to "yesterdays," "September in the Rain" (1937), has a dynamic melody that Dubin matches with vivid imagery and rhythmic phrasing. He may have been drawing on Phil Baxter's "A Faded Summer Love" (1931), with its nostalgic evocation of falling leaves, "Some of them are brown, some are red" (which nevertheless contains a neat "some are"/"summer" rhyme). Given Warren's swinging music, however, Dubin followed suit with "the leaves of brown came tumbling down—remember? in September—in the rain?" The image of driving rain makes both leaves and memory more vivid, a vividness intensified by the pounding phrases that prod memory, such as the slangy intensity of "the sun went out just like a dying ember," not "in" but "*that* September in the rain." By the end of the song he has earned his hyperbolic claim, offhandedly phrased, which makes autumnal memory more vivid than vernal present:

> Though spring is here,
> to me it's still September.

Dubin had such vernacular ease only for Harry Warren's music. When Warren left Warner Brothers for Twentieth Century-Fox in 1938, Dubin reverted to his sentimental style—almost immediately, in fact. In 1939 he set sonorous words to an old Victor Herbert melody, "Indian Summer"—a lugubrious reversal of the brassily nostalgic "September in the Rain."

At Fox, Warren's music inspired another lyricist, Mack Gordon, to match his driving melodies with extended lyrical phrasing, long on vernacular ease, short on showy rhymes and "poetic" imagery. Before teaming with Warren, Gordon had collaborated with composer Harry Revel on songs for over thirty films, all done—Hollywood fashion—without seeing a script. Equally Hollywood was the style—as pure and simple as the short, turn-around phrasing of Freed and Brown or Kalmar and Ruby. Occasionally, however, Gordon enlivened the inanely repetitive melodies of Harry Revel (another glaring omission from Alec Wilder's pantheon) with vernacular flashes, such as the quick comeback to "Did You Ever See a Dream Walking" (1933). Similarly, in "The Loveliness of You" (1937) he enlivened his cliché-ridden catalog—"the beam in your eyes, the smile on your face, the touch of your hand"— with such clever concoctions as "when we're cheek-to-cheeking" and "the heaven-above-liness of you."

Teamed with Harry Warren, however, Mack Gordon, like Dubin before him, showed new resources: not only could he use alliteration and assonance to match Warren's rhythmic punch in songs like "Chattanooga Choo-Choo" (1941) and "I've Got a Gal in Kalamazoo" (1942), but he displayed a lightly comic touch. In "I Had the Craziest Dream" (1942) he created a wryly erotic character and situation:

> I had the craziest dream
> last night—
> yes I did!
> I never dreamt it could be,
> yet there you were—
> in love with me.

The hesitant, yet insistent progression of colloquial phrases is subtly sensuous and builds to a prosaically surreal climax, capped by a deft rhyme:

> I felt your lips close to *mine*,
> so I kissed you—
> and you didn't *min-*
> d it at all!

The easy conversational movement here is perfectly suited to a dream where the "craziest" happening turns out to be the most ordinary kiss.

Even at the height of World War II, when Hollywood was upping the sentimental ante by resurrecting such old tear-jerkers as "As Time Goes By" (1931) for the film *Casablanca* (1942) and "I'll Get By" (1928) for *Follow the Boys* (1944), Mack Gordon insulated Harry Warren melodies with witty, vernacular cool. Although the title phrase of "There Will Never Be Another You" (1942) is an open invitation to purple prose, Gordon creates a comic reversal akin to Edna St. Vincent Millay's use of hyperbole to undercut mawkishness in her love sonnets. Conceding that "there will be many other nights—like this," Gordon's urbane lover follows Warren's ranging melody with digressive speculations,

> other songs to sing,
> another fall, another spring,

before returning to the moment at hand with an understated, "but there will never be another you." Rather than concentrate on the moment at hand, however, this imaginative lover is more intrigued with the future he has conjured up. So taken is he with that vision, rather than the palpable present, he concedes that "there will be other lips that I may kiss," then gets so carried away he speaks of the present memory as a past memory: "but they won't thrill me like yours *used* to do." The beloved so addressed may well wonder if she's more interesting as a sensuous presence or a ghostly memory.

Given an equally open invitation to sentimentality with the title "You'll Never Know" (1943), Gordon performed another comic reversal with hard-edged, vernacular phrases. After allowing his ardent lover to pledge love eternal, he shifts to an exasperated defensiveness:

> you ought to know,
> for haven't I told you so—

Then, as Warren's melody takes a surprising leap, this romantic pledge sounds more like a parental scolding: "—a million or more times?" By the end of the song even the sentimental title phrase is undercut with a street taunt—"you'll never know—if you don't know now."

While Warren's swinging melodies could inspire Gordon's witty turns, when the music became too rangy and driving—what Alec Wilder lovingly terms "groovy"—Gordon fell back upon strained clichés. "Serenade in Blue" (1942) suggests, by its title, that it started out as an instrumental, and Gordon's lyric struggles to toss in references to the music itself. "When I hear that serenade in blue," he starts out as Warren's melody quickly traverses an octave, then, as it casually descends the same interval, Gordon limps along with "I'm somewhere in another world alone," lamely appending "with you," as Warren chromatically drops down another half-step. When the music drops even further, Gordon fumbles along with hackneyed phrases:

> sharing all the joys we used to know
> many moons ago

As with so many of Duke Ellington's jazz melodies that later had awkward lyrics attached to them, "Serenade in Blue" should probably be preserved as a purely instrumental number.

In 1945 Harry Warren changed studios again, from Twentieth Century-Fox to MGM. Teamed with less swinging composers, such as old-timer Jimmy Monaco, Mack Gordon contributed to the glut of wartime treacle with such songs as "I Can't Begin to Tell You" (1945), an echo of "You'll Never Know" but without any of its witty exasperation. Harry Warren collaborated with various lyricists for another fifteen years, though his melodies, exemplified by such songs as "Inamorata" (1955) and "An Affair to Remember" (1957), gradually became as bloated as the general style of the era. During the golden age, however, the lyrics of Al Dubin and Mack Gordon provided an indigenous exception to the general run of Hollywood's mill—a more prosaic, slangy, expansively phrased lyric line that matched the driving, rangy melodies of Harry Warren.

12

Swingy Harlem Tunes: Jazz Lyricists

> Harold will write a tune, and it may go all over the place musically, and then you've got to fit that tune. Very touchy work.
>
> YIP HARBURG

One day in 1925 two Alley tunesmiths, Harry Akst and Joe Young, decided to try a new avenue for plugging their latest song. Rather than seeking to place it in a vaudeville show or a Broadway revue, they went to a nightclub, Salvin's Plantation Club, where they sought out a young black singer named Ethel Waters. When Akst and Young approached her with "Dinah," they demonstrated it by singing in a typically bouncy '20s style. "Is that the way you want to sing it?" Waters asked skeptically. The songwriters had the good sense to urge her to sing it her own way. When Waters gave "Dinah" a torchy blues delivery, the cute rhymes on "finah," "Carolina," and "ocean liner" took on a husky edge, and the pun "Dinah might" made for a smoldering climax. Not only did "Dinah" titillate the nightclub audiences, it went on to become a hit on Tin Pan Alley through sheet music and record sales—the first major popular song to emerge from a nightclub.

Initially billed as "Sweet Mama Stringbean," Ethel Waters was one of a number of black nightclub singers who were attracting white audiences to Harlem. Edmund's Cellar, for example,

244

where she had started out, was a club for blacks, but soon whites began filling its seats. By the time she moved to Rafe's Paradise, she had an exclusively white audience. Like Bessie Smith, Ma Rainey, and other "classic" blues singers, Waters not only sang authentic blues but gave Tin Pan Alley songs a blues rendition. Such flourishes were enough to satisfy Prohibition audiences who came to Harlem nightclubs to drink and to listen to "hot tunes and torch songs."

By the late 1920s Tin Pan Alley had begun taking the commercialized blues and jazz that emanated from Harlem nightclubs and transforming them into popular songs. Probably the most popular song of the golden age, "Star Dust," was introduced at the Cotton Club in 1929. As was typical of most nightclub songs, "Star Dust" had originally been a jazz instrumental and became popular only when a lyric was added. Adding lyrics to such music, however, was extremely difficult, given its melodic range and rhythmic complexity. Most of Duke Ellington's nightclub songs, as we shall see, were saddled with cumbersome lyrics when they made the transition from jazz instrumental to "popular" song.

Yet such music could sometimes inspire a talented lyricist to come up with a sensuously vernacular setting that set nightclub songs apart from theater or film songs of the period. Dorothy Fields, as we have seen, got her start at the Cotton Club with "I Can't Give You Anything But Love"; when she and Jimmy McHugh left for Hollywood in 1930, they were replaced by a new songwriting team, Ted Koehler and Harold Arlen. Koehler quickly proved himself to be one of the few lyricists who could successfully write for music as rhythmically complex and melodically rangy as Arlen's. For the next few years he and Arlen wrote songs at the Cotton Club for such performers as Cab Calloway, Lena Horne, and, most notably, Ethel Waters, which were, according to Jack Burton, "the first night club songs to challenge numbers from Broadway musicals."[1]

Koehler and Arlen were an unusually well-matched songwriting team: both had started as singer-pianists, and that common background helped shape their adaptations of blues and jazz idioms. While Arlen knew and admired the work of Kern and Gershwin, his "beginnings," as he recalled, "were never centered on those guys, only jazz." His main ambition, in fact, was to be a singer. His father was a cantor in a Buffalo synagogue, and Arlen had developed a singing style that blended his Jewish roots with his newfound love of black jazz and blues. Because his father had originally come to America to sing at a Louisville synagogue, Arlen was certain he "must have picked up some of the blacks' inflections." Arlen recalled how once he played a Louis Armstrong record for his father, and at an Armstrong riff ("we used to call it a 'hot lick' ") his father was stunned and demanded to know, in Yiddish, "Where did *he* get it?"[2]

Starting out in Buffalo nightclubs, Arlen, still going by his real name, Hyman Arluck, began to develop a singing style—later adapted by Frank Sinatra—of "hovering around a note and employing a tremolo nuance he had learned in his father's synagogue." One of his great early triumphs came when he was singing with his jazz group, the Buffalodians, in a New York nightclub; after his performance, the legendary Bix Beiderbecke came up to Arlen and said, "Great, kid." "Holy Jesus," Arlen recalled, "that meant so much to me!" In 1929 he had a singing role in Vincent Youmans' musical *Great Day*. One day during rehearsal the pianist, Fletcher Henderson, took ill, and Arlen offered to fill in. As he rehearsed the Will Marion Cook Negro Chorus on the title number, Arlen improvised a piano vamp that he embellished as the chorus went over the routine. Finally, during a break, Will Marion Cook himself went over to the piano and told Arlen, "Boy, you gotta tune there in that little old vamp. And it's a mighty pretty one, too. Better put it down on paper and sell it 'fore somebody steals it off you."[3]

Even though Arlen, by his own admission, knew nothing about composing, he took Cook's advice—and his vamp—to Harry Warren, then an Alley composer on the staff at Remick's publishing house. Warren put him in touch with Ted Koehler,

who matched Arlen's vamp with a driving sequence of slangy
imperative phrases that emphasized the word "get":

> For*get* your troubles
> and just *get* happy,
> *Get* ready for the Judgment Day.

With its liberal sprinklings of "hallelujahs," "Get Happy" sounds
like a typical Tin Pan Alley "gospel" song, but it was successful
enough to keep Koehler and Arlen together.

"Get Happy" also revealed that one way to match Arlen's
wide-ranging melodies and driving rhythms was to build a lyric
around short verb phrases. In "Between the Devil and the Deep
Blue Sea" (1931), for example, the musical accents fall neatly on
the verbs to give the lyric a nervous, frustrated energy.

> I should *hate* you,
> but I *guess* I *love* you,
> you've *got* me in between
> the devil and the deep blue sea

Koehler had also found that alliteration, particularly on such
guttural verbs as "get" and "got," highlighted Arlen's stomping
accents, and he used such alliteration again in "I've Got the
World on a String" (1932):

> I've got
> a song that I sing,
> I can make the rain go

Instead of clever rhymes, which might have made the lyric seem
more "poetic," Koehler relies more on alliteration and simply
repeats "go" at the end of the release (but deftly shifts its meaning from "move" to "release"): "I'd be a silly so-and-so, if I should
ever let go."

He worked similar effects in "I Gotta Right to Sing the Blues"
(1932), a song that comes closer to authentic blues with its twelve-

bar AAB structure. The catch-phrase title sets the tone of the lyric, a blend of southern lament and New York disputatiousness:

> I gotta right to sing the blues,
> I gotta right to feel lowdown,
> I gotta right to hang around,
> down around the river

Again Koehler relies on alliterating *g*'s and *d*'s to follow Arlen's sharp musical contour, and when Arlen makes a blues-like octave-long drop, Koehler follows with a lyrical descent on a richly alliterative "down around the river," ready-made for vocal improvisation. Koehler himself handles the word "around" improvisationally—shifting its meaning from "hang around" through "down around the river" to "dragging' my poor heart around." To blunt the prosaic lament, Koehler chooses faint off-rhymes, such as "down" and "around."

Their biggest hit, the earthy aria "Stormy Weather" (1933), was actually written with Cab Calloway in mind, but after Ethel Waters introduced it, under a lamppost with a blue spotlight on her, the song became her trademark. It was George Gershwin who pointed out to Harold Arlen that he "did not repeat a phrase in the first eight bars"—a nightclub song departure from the repetitive patterns that typified the melodies of the 1920s. Such innovation, according to Alec Wilder, "caused a great deal of consternation in the rhythm section of the bands,"[4] but it inspired Koehler to weave an extended blues-like lament out of truncated, slang phrases:

> don't know why
> there's no sun up in the sky,
> stormy weather,
> since my man and I
> ain't together,
> keeps rainin' all the time

While not one line is actually repeated, the omission of pronouns and connectives gives the lyric a nervous monotony (akin to Berlin's bluesy "Supper Time," also written in 1933). Koehler intensifies that relentless feel with *i*-rhymes that fall as incessantly as the rain; to keep them going he even opts for proper grammar ("my man and I" instead of the slangier "my man and me") and splits "time" over two notes to elongate the *i*. When he does turn ungrammatical with "ain't," he makes it do double-duty by rhyming internally with "*rai*ning."

The slangy rhymes spill over into the vernacular complaint, "Just can't get my poor self together," and Koehler gives "together" his usual improvisational twist by changing it from its earlier meaning in "my man and I ain't together." He comes even closer to blues lyrics, with their wry imagery, in the release, where he personifies "the blues":

> when he went away
> the blues walked in and met me,
> if he stays away
> old rockin' chair will get me

No sooner is that comic touch introduced, however, than the lyric swerves upward with Arlen's music in a soaring plea to "walk in the sun once more," a plea subtly underscored by having "walk" echo "*rock*in' chair" and intensified by the back-to-back rhyme "*sun*/*once*." Even more wrenching is the off-rhyme that opens the final section with the forlorn lament: "can't *go on,* ev'rything I have is *gone.*"

Koehler and Arlen also had a rhythmic hit in 1933, "Let's Fall in Love," though it was written for a movie rather than the Cotton Club. Still, it has the same brassy pugnacity of their nightclub songs, as Koehler turns the clichés of romantic proposal into an aggressive sales pitch with rhetorical questions—

> Why shouldn't we fall in love?

urgent deal-clinchers—

> Let's take a chance

and dangling bargains—

> We might have been meant for each other

Ted Koehler and Harold Arlen were clearly meant for each other
in such songs, but it was Arlen's desire to break out of the
"groove" of blues and swing ("for fear of being typed"[5]) that
ended their collaboration at the Cotton Club in 1934. When he
shifted from nightclub songs to Broadway revues, Arlen also
changed lyricists, working with Yip Harburg on such satiric
shows as *Life Begins at 8:40*. When he moved from Broadway to
Hollywood, Arlen worked with various film lyricists, most notably Johnny Mercer, with whom he recaptured his blues style.

Koehler, too, went to Hollywood, where he worked with
various composers, but he never really established the same kind
of rapport he had had with Arlen at the Cotton Club. He had a
rhythmic hit, "I'm Shooting High" with Jimmy McHugh in 1935
and a successful ballad, "What Are You Doing the Rest of Your
Life," with Burton Lane in 1944, but he frequently capitulated to
Hollywood's standardized fare, as in "Some Sunday Morning,"
where he set M. K. Jerome's bland melody to a lyric that smacked
of "For Me and My Gal," right down to such thinly disguised
borrowings as "Bells will be chiming an old melody, 'spec'lly for
someone and me."

Ted Koehler was one of the few lyricists who could skillfully set
music that used the extended, driving phrases of jazz and blues,
and Harold Arlen was more fortunate than most such composers
in having not only Koehler but, as we shall see in the next chapter, Johnny Mercer as his collaborator. Given the dearth of lyricists as skilled as Koehler, some composers simply put their own
words to their instrumentals. Such a tactic had an added advantage for a woman songwriter like Ann Ronell, and she further
bypassed the old-boy network of Tin Pan Alley by selling her
song, "Willow Weep for Me" (1932), directly to a bandleader,
Paul Whiteman. Although she dedicated it to George Gershwin,

the melody has more Arlen in it, and her languorously driving lyric is much closer to Koehler than to Ira Gershwin:

> Oh weeping willow tree,
> weep in sympathy,
> bend your branches down
> along the ground
> and cover me,
> when the shadows fall,
> bend, oh, willow and
> weep for me.

Like Koehler, Ronell keeps her lyric prosaic with simple repetitions (*weep*, *me*, *sympathy*) and off-rhymes (*down/ground*), relying more upon alliterating *w*'s and *l*'s to emphasize the incessant imperatives. Image and phrasing here blend as the skillful extension of syntax keeps the lament tumbling, willow-like, forward and downward with the rangy music.

Another composer, the Englishman Ray Noble, wrote his own words for "The Very Thought of You" (1934), and also used faint rhymes and alliteration to anchor a syntactically long and driving lyrical line:

> The very thought of you
> and I forget to do
> the little ordinary things
> that ev'ryone ought to do

The colloquial ease here stems from his title, which transforms a simple catch-phrase from its ordinary context of disdain (the very thought of it!) to a sensuous complaint. Throughout these vernacular phrases Noble weaves subtle repetitions—rhyming *very* with ordin*ary*, then embedding it in "*ev'ry*one." When he had to come up with a parallel phrase for his title, he hit upon "The mere idea of you," which carries the *er* from "very" over into "mere." Equally deft is his double rhyme on "daydream" and "may seem," which further drives the lyric forward with the melodic line.

Even when a lyricist and composer did manage to collaborate successfully on a blues song, it was usually a one-time affair.
One of the decade's great torch songs, "You Go to My Head"
(1938), was written by the unlikely-sounding team of Haven Gillespie and J. Fred Coots (their only other hit together was "Santa
Claus Is Coming to Town"). Originally written in 1936, "You Go
to My Head" was turned down by one Alley publisher after another, until in 1938 Glen Gray's orchestra made it a hit. An
"introspective song" that encourages singers "to use their most
personal and intimate approach,"[6] "You Go to My Head" soon
became a standard part of the repertoire of Billie Holiday, Lena
Horne, Peggy Lee, and other singers who gave it the full-flame
treatment.

Alec Wilder terms Coots' melody a "minor masterpiece,"
and Gillespie's lyric not only matches it with smoothly expanding
phrases but artfully weaves an elaborate skein of imagery. Reaching back to Berlin's comparison of a "pretty girl" to the "haunting refrain" of a recurring melody, Gillespie updates the metaphor by invoking the image of a phonograph record:

>   you go to my head
>   and you linger like a haunting refrain
>   and I find you spinning 'round in my brain

Then he introduces a strand of the alcoholic imagery implicit in
the catch-phrase title:

>   like the bubbles in a glass of champagne

Gillespie continues the alcoholic imagery in the next section, but
he comes down a notch in vintage elegance with "You go to my
head like a sip of sparkling Burgundy brew." He then modulates
the imagery further down the scale with a slangy American
potable:

>   and I find the very mention of you
>   like the kicker in a julep or two

Although he drops the alcoholic imagery in the release, he recaps it subtly in the final A section by reminding us that alcohol, like mercury, rises in thermometers:

> you go to my head
> with a smile that makes my temp'rature rise,
> like a summer with a thousand Julys,
> you intoxicate my soul with your eyes

Rather than adopt the list strategy of Cole Porter by trying to "top" each successive image, Gillespie seems to abandon the alcoholic metaphors altogether, only to slip in "intoxicate," which closes the sequence with a kicker. Although the song echoes Porter's "I Get a Kick Out of You," it has its own sensuous and languorous sophistication.

Another successful one-time collaboration, this time between Johnny Burke and Bob Haggart, produced "What's New?" (1939), which had started out as an instrumental solo for trumpeter Billy Butterfield, until, according to Alec Wilder, Burke's "wonderfully conversational lyrics"[7] transfigured it into a popular hit:

> what's new?
> probably I'm boring you—
> but seeing you is grand—
> and—
> you were sweet—
> to offer your hand

What's new about this lyric is that it is a full-fledged dramatic monologue, one side of a conversation between two ex-lovers who meet by chance on the street. Burke shrewdly used the soaring, wide-ranging melody as a counterpoint to the lovers' low-keyed but strained small talk. The tension between music and lyric reveals one lover's unspoken ardor, right down to his nervously cheery "adieu." Only the prosaic "I haven't changed" works as a pun that finally releases his suppressed (but barely whispered) "I still love you so" as the music trails downward.

Another dramatic monologue whose chatty lyrical phrases are counterpointed against a sinuous melody was Earl Brent and Matt Dennis' "Angel Eyes" (1946), a song tailor-made for night-club torch singers. The lyric was made even more dramatic by Frank Sinatra, who began his rendition at the release, which sets the scene in a crowded bar:

> so drink up, all you people—
> order anything you see—
> have fun—
> you happy people,
> the drink—
> and the laugh's—
> on me

By returning to the release at the end of the song, Sinatra gave the final A section the denouement of a miniature play:

> pardon me,
> but I gotta run—
> the fact's uncommonly clear—
> I gotta find
> who's now number one—
> and why
> my angel eyes ain't here—
> 'scuse me—
> while I disappear.

With its tension between driving music and restrained lyrics, "Angel Eyes" is in the tradition of the greatest of all torch songs—Mercer and Arlen's "One for My Baby" (1941).

The brief collaborations of Ted Koehler and Harold Arlen and the even briefer one-night stands of teams like Gillespie and Coots were, nonetheless, genuine *collaborations,* where lyricist and composer worked together to complete a song. Much of the poor quality of lyrics set to jazz and blues stems from the fact that the lyricist was frequently called in after, sometimes years after, the music was completed. In some cases there was no real collaboration whatsoever; the lyricist's task was simply one of setting words to an already composed (and sometimes already popular) jazz instrumental. The most prolific of such wordsmiths was Mitchell Parish, who worked as a staff lyricist for the Irving Mills music publishing company, a company that specialized in transforming jazz instrumentals into mainstream pop songs by adding a lyric. While Parish's lyrics frequently effected such a transformation, his language strove for elevated heights rather than vernacular ease. Thus, though he could weave syntactically driving phrases to match the difficult melodies he was given, the diction and imagery of such phrases are poetically overwrought.

In 1928 Parish was assigned to set a lyric to the instrumental "Sweet Lorraine," but the following year he had his biggest hit, and probably the most popular song ever to come out of Tin Pan Alley, "Star Dust." Hoagy Carmichael had written "Star Dust" in 1927, when he was at the University of Indiana—supposedly while sitting on the "spooning wall" one night and recalling an old flame. Former classmate Stu Correll dubbed it "Star Dust" because "it sounded like dust from stars drifting down through the summer sky"[8] (though he may have had an assist from *Poetry* magazine, which published a selection of Emily Dickinson poems under the title "Star Dust.")

Originally a piano rag, "Star Dust" was recorded as an upbeat instrumental by Don Redman and his orchestra, but then an arranger suggested that it "be played in slower tempo and in a sentimental style." Once the pace was slowed and the nostalgia thickened, it occurred to publisher Irving Mills that it could be a commercial success, and Mitchell Parish was called in. The task Parish faced with "Star Dust" was a difficult one: as Wilder notes, the song's "instrumental beginnings are obvious," its "melodic line is not, in the pure sense of the word, vocal," but "unconventional" and "not easy to sing."[9] Parish's skill is manifest in the way he created a long, but conversational line that followed the contour of the melody, pausing with the music, but then driving forward syntactically:

> Sometimes I wonder why
> I spent the lonely night
> dreaming of a song?
> The melody
> haunts my reverie,
> and I am once again with you,
> when our love was new,
> and each kiss an inspiration,
> but that was long ago:
> now my consolation
> is in the star dust of a song

As long as that sentence is, its phrases tumble into one another as effortlessly as Carmichael's hero, Bix Beiderbecke, might improvise the phrases of a cornet solo. Parish shrewdly made the subject of the song the melody itself—its drifting, delicately turning phrases suggesting the metaphor of the title.

Yet while Parish's phrasing is utterly casual, his imagery and diction strain to be "poetic," and the lyric gathers up all the dusty trappings of the Alley—bright stars, nightingales, fairy tales, and even doubles up clichés like "paradise where roses grew." Still more high-pitched was the lyric Parish wrote when Carmichael added a verse to make "Star Dust" conform even more to the pop song format. Once again, the phrasing and theme are perfectly matched to the musical drift: dusk "steals," stars "climb," and, varying the sequence of passages, "you wandered down the lane and far away." Yet, again, diction and imagery turn the passage purple: "a song that will not die" is "the music of the years gone by," the stars are "little" and climb "high up in the sky," and the dusk that "steals across the meadows of my heart" is, appropriately, "purple."

Even though it was a far cry from the vernacular lyrics of Dorothy Fields and Ted Koehler, "Star Dust" became an immediate hit after it was introduced at the Cotton Club in 1929, and its purple prose colored Parish's other lyrics as well. In "Take Me in Your Arms" (1932) he pleaded, "Blind me with your charms—with all the star dust in the sky." In "Stars Fell on Alabama" (1934) Parish upped the astral ante from a fall-out of mere dust to the stars themselves. So reliant was he upon the imagery of "Star Dust" that ten years later, when he was called in to set an instrumental by Peter DeRose (originally written as a piano piece in 1934), he made "Deep Purple" a dust that "falls over sleepy garden walls."

Like most wordsmiths who set lyrics to instrumentals, Parish never established an enduring collaboration with a composer. He was one of over a dozen lyricists, for example, who at one time or another was called in by Irving Mills to put words to Duke Ellington's songs. (Mills became Ellington's manager in 1926 and is credited with the band's move from the Kentucky Club to the Cotton Club and Ellington's subsequent rise to fame.) The unfortunate result of such one-time assignments is that the superb songs of Duke Ellington seldom received good lyrics. Parish's sentimentally didactic lyric for "Sophisticated Lady," for example, actually works *against* Ellington's elegantly sensuous music. Although Parish's syntax, as always, skillfully drives forward with the rangy melody, his lyric tells a moral tale that goes all the way back to "A Bird in a Gilded Cage" (1900), replete with the same poetic images and inversions:

> They say into your early life romance came,
> and in this heart of yours burned a flame,
> a flame that flickered one day and died away.
> Then, with disillusion deep in your eyes,
> you learned that fools in love soon grow wise

When Parish gets to the chorus his lyric becomes "excruciating," in Alec Wilder's judgment, "scarcely lyrical,"[10] though, once again, the structure of the phrasing is perfectly conversational:

> smoking, drinking,
> never thinking

> of tomorrow,
> nonchalant,
> diamonds shining,
> dancing, dining
> with some man in a restaurant,
> is that all you really want?

Parish's puritanical finger shakes not only at the sophisticated lady's smoking, drinking, and dancing but at her "nonchalant" style—the style of Ellington's music itself.

One can hardly blame jazz critics for ignoring the lyrics that were added to Duke Ellington's instrumentals, since Ellington was rarely blessed with lyricists who matched his music with words that were both slangy and sophisticated. At least his earliest popular songs, from the Cotton Club shows of the late 1920s, had lyrics that tried to be vernacular. Working with singer Cab Calloway, Irving Mills and Clarence Gaskill came up with a tale of "Minnie the Moocher" that included such earthy vignettes as that of "Old Deacon Low-down," who

> preached to her she ought to slow down
> but Minnie wiggled her jelly roll
> Deacon Low-down hollered, "Oh, save my soul!"

When Mills tried to go it alone, however, he took Ellington's marvelously slangy title, "It Don't Mean a Thing If It Ain't Got That Swing" (1932), and followed it with a plethora of "doo-wahs" and lines that sound precisely like a white lyricist trying to sound blackly "hip":

> It makes no diff'rence if it's sweet or hot,
> just give that rhythm ev'rything you got

Perhaps inspired by Mitchell Parish, Mills and company soon shifted from stilted slang to soaring "poetry." With Eddie DeLange, Mills sprinkled star dust on "Solitude" (1934): "you haunt me with reveries of days gone by" and "memories that never die." A year later, for Ellington's wry "In a Sentimental Mood," Mills and Manny Kurtz strained for the heights:

on the wings of ev'ry kiss
drifts a melody, so strange and sweet,
in this sentimental bliss
you make my paradise complete

Alec Wilder wonders about Mills' co-conspirator here:

> And who, may I ask is Manny Kurtz? We know by now of the
> omnipresence of Irving Mills on the credits of Ellington
> songs . . . I'm bemused! Mr. Kurtz is obviously the lyric writer.
> And, frankly, only by the skin of his teeth, as the words cer-
> tainly don't fall very fluidly.[11]

Mills, working alone, could produce even more melodramatic—
and illogical—astral displays. "Caravan" (1937), for example, begins with "stars above that shine so bright" then suddenly
switches to the "myst'ry of their fading light."

Occasionally, however, the teams at Mills could come up with
lyrical phrasing more suited to Ellington's wide-ranging melodies. Teamed with Albany Bigard, Mills came up with a languorously melancholy lament for "Mood Indigo" (1931):

You ain't been blue
no, no, no—
you ain't been blue,
till you've had that mood indigo

Taking the elegantly inverted title (not "indigo mood" but "mood
indigo") they wedded it not to sonorous poetics but to driving
conversational phrasing, accenting the forward movement with
insistent rhymes—indigo echoing *no, blue* and *you*'ve picking up
the long vowel of *mood*.

Again in 1938 Mills and two fellow lyricists, Johnny Redmond and the suspiciously named Henry Nemo ("Nemo," we
recall, was the Greek word for "No man," which Odysseus used
as a pseudonym to fool the Cyclops) lamented with obvious
strain, "I let a song out of my heart—it was the sweetest
melody—I know I lost heaven—'cause you were the song," and
pleaded

> Am I too late to make amends?
> You know that we were meant
> to be more than just friends

Despite the deft off-rhyme on *meant* and *amends*, the terms of endearment here hardly rise above greeting-card verse.

It was in the 1940s, after Ellington ended his association with Mills, that he finally found lyricists who could handle the vernacular skillfully enough to match his music. Paul Francis Webster, who later wrote such bloated paeans as "Love Is a Many-Splendored Thing" (1955) and "A Very Precious Love" (1958), managed to weave colloquial catch-phrases together in a way that fleshed out the wonderfully wry title of "I Got It Bad and That Ain't Good" (1941). The chorus opens with clipped, telegraphic phrases,

> Never treats me
> sweet and gentle
> the way he should

Even when he ventures toward the maudlin, as in "my poor heart is sentimental," Webster brings the lyric back down to earth with the casual afterthought "not made of wood." The best vernacular touch comes in the release, where the standard motif of the lover trapped on the treadmill of fate takes a slangy turn:

> But when the weekend's over
> and Monday rolls aroun'
> I end up
> like I start out
> just cryin'
> my heart out

The deft handling of slang here sets up the intricately ungrammatical "He don't love me like I love him—nobody could" that closes off the song with driving, but resigned, negations.

Webster's telegraphic slang may have set an example for Bob Russell, who put a lyric to Ellington's "Don't Get Around Much Anymore" in 1942. What Alec Wilder terms the "daring" opening line—"Missed the Saturday dance"—recalls Webster's dropping of pronouns, and Russell tersely omits his *I*'s throughout the lyric:

> —heard they crowded the floor
> —couldn't bear it without you
> —don't get around much anymore

The clipped understatements disclose feeling, even as the slang gets more pugnacious: "Been invited on dates—might have gone—but what for?" With a throwaway line like "Awf'lly diff'rent without you," Russell gets more genuine sentiment out of an Ellington melody than Irving Mills and all of his collaborators rolled together. He did the same in "Do Nothin' Till You Hear From Me" (1943)—probably the slangiest pledge of romantic fidelity ever written—and "I Didn't Know About You" (1944) with its disputatiously affectionate protestation: "How could I know about love—I didn't know about you?"

Ellington's luck with lyricists hit its peak in 1944 when Don George gave an exuberantly casual setting to the catch-phrase "I'm Beginning to See the Light." Still, according to George himself, it was difficult to interest anybody—even Johnny Mercer—in the lyric. Ellington's tune was particularly hard to set, since each A section consists of the same, driving vamp-like phrase repeated three times over before the melody finally changes. In one way, George heightened this musical insistence, using the same rhyme for the first three lines of each section.

> I never cared much for moonlit skies,
> I never wink back at fireflies,
> but now that the stars are in your eyes,
> I'm beginning to see the light.

Yet, as we can also see here, he set up a structural counterpoint to the musical monotony: while the first two lyrical phrases are parallel, the third changes syntax, then in the fourth line, when the music finally changes, George repeats his title phrase.

What is even more artful about the lyric is George's witty use of a list structure; not only does he develop a catalog of "light" images, he refreshes the tritest metaphors by reading them in literal terms—from stars in the eyes, to a suggestive "afterglow," and the paradoxical "now when you turn the lamp down low I'm beginning to see the light." It's in the release, however, that his literal metaphors really shine:

> Used to ramble thru the park
> shadow-boxing in the dark
> then you came and caused a spark,
> that's a four-alarm fire now

George ends by rekindling one of the oldest songwriting clichés, mixing his metaphors of light and heat: "but now that your lips are burning mine, I'm beginning to see the light."

No sooner had lyricists such as Webster, Russell, and George shown what could be done with Ellington's music that his rangy melodies fell once again into the hands of Irving Mills, who began turning out such strains as "I Don't Know Why I Love You So" (1947) and "Indigo Echoes" (1947). It is ironic that during the golden age of popular song, Duke Ellington, the best jazz composer of the era, seldom had his instrumentals set by lyricists with even the slightest flair for his vernacular elegance. That ironic point was driven home when an Ellington instrumental that seemed too marvelous for words, "Satin Doll," was given a perfect lyric by a writer who, in the best of all possible worlds, would have been Ellington's exclusive collaborator—Johnny Mercer.

13

Midnight Sun:
Johnny Mercer

*I can remember liking Kern's "They Didn't Believe Me"
when it came out. Couldn't have been more than four or
five then.*

JOHNNY MERCER

After he won the Academy award in 1942 for "The Last Time I
Saw Paris," Oscar Hammerstein whispered to a friend, "If you
see Johnny Mercer, tell him he was robbed." Mercer and Arlen's
"Blues in the Night" had also been nominated for best song, but
it was the sentimental tribute to Nazi-occupied Paris—"No matter how they change her, I'll remember her that way"—that won
out over Mercer's unpitying lament:

> A woman'll sweet talk,
> and give ya the big eye,
> but when the sweet talkin's done,
> a woman's a two-face,
> a worrisome thing who'll leave ya t'sing,
> the blues in the night

Mercer *was* robbed—and not just of an Oscar by an Oscar. Although he broke into songwriting at the height of the golden
age, by the time he hit full stride, the era was over.

After *Oklahoma!*, Broadway began closing its doors to lyricists like Mercer, who "thought in terms of the song rather than the show."[1] Hollywood, too, made fewer and fewer original musicals and could offer Mercer only the limited (but lucrative) piecework of setting words to movie themes, like "Moon River" (1961) and "Days of Wine and Roses" (1962). Even the straight pop market was changing; Tin Pan Alley not only disappeared as the center of the musical universe, but it was replaced by the black and country music it had once carefully relegated to the peripheries as "race" and "hillbilly" records.

Mercer's roots, ironically, were in such music. Even at his most urbane, Mercer was always the country boy from Savannah—even more of an anomaly on the Alley than Cole Porter. He had learned his trade by listening not only to the songs of Kern and Berlin but also to the "race" records marketed to southern blacks. It was those roots, further enhanced by his talent as a singer, that made Mercer one of the few lyricists who could skillfully set the jazz melodies of composers such as Hoagy Carmichael, Harold Arlen, and Duke Ellington. What gives Mercer's best songs their distinctive character is their blend of urbanity and earthiness, a blend so distinctive that, alone among the lyricists of Tin Pan Alley, people speak of a "Mercer" song as readily as they denominate songs by their composer.

Not all Mercer songs, not even all of his hits, manifest that marriage of Manhattan and Savannah. In his early lyrics, written for Paul Whiteman's band, where Mercer worked as a singer-songwriter, we can hear various styles clashing. His first success, "Out of Breath" (1930), starts off with the witty rhymes of his earliest idol, W. S. Gilbert—"When tasks super*human* demand such a*cumen* that only a *few men* possess"—but then suddenly shifts to the American vernacular:

> All you have to do
> is say "Boo!"
> Out of breath
> and scared to death
> of you.

Even more stylistically confusing is "Would'ja for a Big Red Apple" (1932), whose country slang is bracketed by lines reminiscent of another early Mercer love, Victor Herbert operettas—"Greater men than I have sought your favor"—alongside still others that smack of Cole Porter's naughtiness—"you spurn the vermin who offer ermine."

Although Mercer clearly longed to write in his regional vernacular, Tin Pan Alley had long relegated that idiom to comedy songs. Teamed with another country boy, Hoagy Carmichael, Mercer wrote the lyric for "Lazybones" (1933) as a protest against all the artifical "Dixies," "Swanees," and "Carolinas" concocted by songwriters who had never ventured further from Manhattan than Brooklyn. Right off the bat it announces its authenticity with "Long as there is chicken gravy on your rice" and asks such homespun questions as "How you 'spec' to get your cornmeal made?" Such regionalism was fine for laughs, but when he tried to use it for love songs, he had to apologize, as in "Pardon My Southern Accent" (1934):

> It may sound funny,
> ah, but honey! I love y'all

For years Mercer had to declare "serious" love in the Alley's standard terms, avowing in songs like "Spring Is in My Heart Again" that "you are here" so "skies are clear."

Surprisingly, it was when Mercer went to Hollywood, in 1935, that his lyrics displayed more sophisticated wit and more citified slang. When he met Fred Astaire in a Hollywood recording studio, the unlikely pair decided to collaborate on a song. "I'm Building Up to an Awful Let-Down" was Mercer's first successful lyric in a casually urbane style, and it seems clear that Astaire, who embodied that style more than any other singer of the era, was its inspiration. While Mercer's lyric practically alludes to Astaire in lines like "my smile so debonair," he gives urbanity his own driving, punchy tang:

My one big love affair—
is it just a flash,
will it all go smash,
like the nineteen twenty-nine market crash?

Here in the telegraphic syntax, the slangy exuberance, and what
Bob Bach aptly terms the "nonchalant bravado"[2] of the stock-
market simile, Mercer found his own distinctively jazzy version of
the colloquial elegance of Hart, Gershwin, and Porter.

Although he kept his southern drawl for such comic songs as
"I'm an Old Cowhand" (1936), Mercer contributed to the lustre
of the golden age with film songs that bristled with wit and vernacular energy. That he was able to do so in Hollywood probably
reflects that fact that there he worked with composers like Richard Whiting and Harry Warren, whose music was more rhythmically driving and melodically rangy than the rickey-tick fare of
most film music. In 1937 he turned out a lyric for Richard Whiting's "Too Marvelous for Words," a song Alec Wilder praises as a
"model of pop song writing," one that departs from "compla-
cency and pedestrianism"[3] with surprising shifts of rhythm and
key. To set such a Gershwinesque tune, Mercer borrowed some
of the techniques of Ira Gershwin; not only does he allude directly to " 'S Wonderful" with "that old standby" rhyme of "glamorous" and "amorous" (hardly a clichéd rhyme, of course), but,
as Gershwin so often did, he writes more about language than
love. In this case, it is the language of love songs themselves that
bedevils him, and Mercer seems to give vent to his own frustration in having to follow such masters as Hart, Gershwin, and
Porter:

It's *all*
too wonder*ful*
I'll
never find the words,
that say enough,
tell enough,

Yet just as he seems at the end of his rope, Mercer's own slangy punch comes through with "I mean, they just aren't swell enough."

Using another Gershwin device, substituting one part of speech for another, Mercer mimes the affected terms of endearment of urban sophisticates:

> you're much too much,
> and just too very very!

The syntactic play of having adverbs and adjectives modify one another testifies to the paucity not only of spoken language but of the whole written reservoir of words: "to ever be in Webster's Dictionary." Mercer repeats sounds—*ever/very, very/dictionary*—to keep the lyric driving forward, and at the end of the release he slides into the final A section with "And so, I'm borrowing a love song from the birds." That borrowed song reworks the slang cliché "for the birds" into a casually understated compliment, cleverly confessing the wordless superiority of nature over verbal artifice.

Mercer also teamed with another of Hollywood's top composers, Harry Warren, whose driving melodies provided a perfect setting for turnaround slang phrasing. When Warren's music for "You Must Have Been a Beautiful Baby" (1938) took a sudden, surprising leap, Mercer followed with the vernacular comeback, " 'Cause, baby, look at you now." Hollywood in the 1930s was a melting pot for American slang, and Mercer kept his ears peeled; when he heard Henry Fonda exclaim "Jeepers creepers" in his Midwestern twang, Mercer latched on to it and added other regionalisms like "Gosh all git up!" Juxtaposed to this yokelese, however, are such citified expressions as "How'd they get so lit up" and "Oh! Those weepers!" With everything from an archaic "Woe" to the latest slang for spectacles, the lyric becomes a collage of clashing idioms:

> Golly gee!
> When you turn those heaters on,

woe is me!
Got to put my cheaters on

In "Hooray for Hollywood," Mercer went even further, making Hollywood slang itself his subject:

That screwy bally hooey Hollywood,
where any office boy or young mechanic
can be a panic,
with just a good looking pan

While he mingles discordant allusions with the skill of Cole Porter—Shirley Temple, Aimee Semple, Donald Duck—Mercer's list foregrounds the very argot of Hollywood, "where you're terrific if you're even good," "a good looking pan," and the Hedda Hopperese of "They come from Chilicothes and Paducahs with their bazookas to get their names up in lights."

Along with his Hollywood duties, Mercer continued writing for bands, though in the mid-'30s he switched from Paul Whiteman to Benny Goodman. Setting words to jazz instrumentals could sometimes bring out Mercer's lyrical vices—his penchant for cuteness in "Goody Goody" (1936), or his "poetic" streak in "And the Angels Sing," fraught with "Silver waves that break on some undiscovered shore" and "long winter nights with the candles gleaming." Given the right composer, however, Mercer could hit his urbanely earthy stride. In 1939, for example, he collaborated with Rube Bloom on "Day In—Day Out," which Alec Wilder, who did the original arrangement, calls "an absolutely great song."

I hadn't seen the song before and was astounded by both the melody and the lyric by Johnny Mercer. It was . . . unlike any song in the pop field I'd ever seen . . . fifty-six measures long. The melodic line soared and moved across the page like a lovely brush stroke. It never knotted itself up in cleverness or pretentiousness. And it had, remarkable for any pop song, passion.[4]

Mercer captured that passion in his lyric by playing with the Porterish paradox of bored excitement. The catch-phrase "day in—day out" implies excruciating routine, as does the "same old . . ." formula, but Mercer contrasts those phrases with exotic and exciting images (much as he later would do by juxtaposing "that old" with "black magic"):

> Day in—day out,
> the same old hoodoo follows me about.
> The same old pounding in my heart whenever I think of you
> and, darling, I think of you
> day in—and day out

Given an unusually long refrain—and no verse—Mercer skillfully weaves a syntactically driving line of appositions by carrying over fragments, "same old" and "think of you," from line to line, mirroring the incessant repetition that is his theme.

Mirroring, in fact, is an apt description of how the lyrical and musical phrases work in this song. The second A section, for example, opens with a reversal of the title phrase: "Day out—day in." Such mirroring underscores the agitated monotony of love, but Mercer keeps his lover desperately nonchalant with "I needn't tell you how my days begin." He keeps the ennui up, building tension underneath it, until both music and lyric are ready to explode:

> Then I kiss your lips
> and the pounding becomes
> the ocean's roar, a thousand drums

Even after the climactic kiss, Bloom's music delays that explosion with back-to-back triplets, which Mercer matches with the nervously jaunty "and-the-poun/ding-be-comes." That pounding goes all the way back to the heartbeat of the opening, a beat at once monotonous and exciting, which then turns into the equally incessant metaphors of "ocean's roar" and "a thousand drums."

When the long-delayed climax does come, it provides a great match of word to music. After "unexpectedly" stretching his phrasing for four measures, Rube Bloom "returns in romantic desperation to the high *F*," and Mercer finds the "perfect" *F* word with the "can't" that introduces the phrase "CAN'T you see it's love?" With that utterly colloquial demand, Mercer releases the repressed energy of the whole lyric and quickly follows up with exasperated catch-phrases that seem to grope for words:

> Can there be any doubt,
> When there it is
> day in—day out

The poetry here lies, as it so often does in Gershwin's lyrics, in having the smallest and least significant bits of speech—"it" and "is," "in" and "out," and the deftly turned meanings of "there"— carry all the romantic weight.

It's difficult to believe that a year later the same team of Mercer and Bloom could come up with "Fools Rush In," (1940). Perhaps it was the proverbial solemnity of the title that induced Mercer to strike a philosophical pose, replete with such sentimental profundities as "And so I come to you, my love, my heart above my head." Even another clever reworking of "there" gets lost in the general stridency:

> Though I see
> the danger *there*,
> if *there's* a chance for me
> then I don't care

What Mercer so obviously needed was a collaborator who could provide him with a steady diet of music at once urbane and earthy.

In 1941 he found him—in Harold Arlen. Arlen's adaptation of jazz and blues provided Mercer with a perfect setting for his slangy sophistication, and for the next few years, in the twilight of the golden age, they collaborated on some of its greatest standards. Their first hit, "Blues in the Night" (1941), recalled the bluesy songs Arlen had written a decade earlier with Ted Koehler at the Cotton Club. Written for a movie originally entitled *Hot*

Nocturne and sung by a black man in a jail cell, "Blues in the Night" was so good the producers changed the title of the movie to fit the song.

The lyric did not come easily, however. Arlen recalls that Mercer "had lots of phrases and lines written down but none of them seemed to fit that opening phrase right. But then I saw those words, 'my momma done tol' me,' way down at the bottom of the pile and I said, 'Why don't we move them up to the top?' It sure worked."[5] The line itself was a penciled substitution for the sentimental and weakly alliterative "I'm heavy in my heart," which opened the original lyric. Arlen's instincts as a singer made him spurn "I'm heavy in my heart" and fasten instead on the line further down, whose long vowels, imbedded in alliterative *m*'s and *d*'s and *t*'s, made a perfect match for his bitterly mournful opening. The shift from "I'm heavy in my heart" to "My momma done tol' me" launched Mercer full-swing into the vernacular phrasing he handled so well.

Like "St. Louis Blues," "Blues in the Night" opens with a twelve-bar musical phrase reminiscent of authentic blues, and Mercer follows suit with some bluesy lyrical repetitions:

> My momma done tol' me
> when I was in knee pants,
> my momma done tol' me—
> "Son—

The abrupt drop on " 'Son' " gives weight to the maternal voice, and Mercer's penciled revisions made that voice more and more dominant. Although the contraction, "a woman'll sweet talk," was in the original lyric, it was followed by "a woman'll glad eye," which Mercer's pencil blackened for the more prosaically biting: "and give you the big eye." He also cut the limp "But pretty soon you'll find" and substituted "But when the sweet-talkin's done" (deftly echoing the opening "done tol' me"). He also followed "a woman's a two-face" with the colloquial "a worrisome thing," instead of the flat "a changeable thing."

Just as the idea of "mirroring" informed "Day In—Day Out,"

here Mercer's strategy of "echoing" runs through the lyric—an echoing that reflects the image of the train's "clickety-clack's a-echoin' back the blues in the night." The black singer, too, has been echoing—"re-calling"—his mother's voice, and when Mercer returns to the "Now" of the present, he hears nature resounding as well. The "rain's a-fallin'," and, as it does, it echoes the "a whooee-duh-whooee" of the whistle "blowin' cross the trestle" (with an off-rhyme on "whistle" and "trestle"). Even the verbs reverberate—the mother's warning of what a "woman'll" do is echoed by such contractions "breeze'll start," "the moon'll hide," and "the mockin' bird'll sing." When Arlen's music returns to the opening phrase, Mercer, too, makes his final phrases resonate with his lyrical beginning—"big eye" and "sweet talk" from the opening maternal counsel are reprised now by the little-boy-grown-up's "I been in some big towns an' heard me some big talk." So too, the powerful vowels and consonants that first caught Arlen's eye in "My momma done tol' me when I was in knee pants" come back in the sounds that open the final section:

> From Natchez to Mobile,
> from Memphis to St. Joe

One can only wonder, with Alec Wilder, "Where, oh where, did Mr. Mercer find . . . *any* of his indigenous, salty, earthy, regional, place-name lines."[6] Mercer, the urbane lyricist with country roots, might answer from the lyric itself: "I been in some big towns an' heard me some big talk."

From the blues, Mercer and Arlen turned to a song in what might be termed the "witchcraft" genre, a genre Mercer himself acknowledged as stemming from Porter's "do do that voodoo." In "That Old Black Magic" (1942) Mercer adapted his paradoxical formula from "Day In—Day Out," combining the routine—"That old"—with the exotic—"black magic." Such paradoxes run throughout the imagery: "those icy fingers up and down my spine" produce "such a burning desire" that "only your kiss can put out the fire." Arlen's music is rife with repeated notes and octave drops, and Mercer's phrasing makes the lyric mime the

musical motion "down and down" and " 'round and 'round." He
starts the second A section, for example, with "The same old
tingle that I feel inside," but then we plunge into a strange new
clause—made all the stranger by the familiar "that": "and then
that elevator starts its ride." But just as suddenly *that* metaphor is
dropped for a spinning simile:

> like a leaf
> *that's* caught in the tide.

In "One for My Baby (and One More for the Road)" (1943)
Mercer and Arlen rang their changes on yet another Alley standard model, the torch song. Mercer presents the lyric as a dramatic monologue set, like "Angel Eyes," in a bar, but rather than
a crowded bar, Mercer creates a deserted scene between the solitary singer and the bartender. Opening with the utterly prosaic
"It's quarter to three," Mercer gives what might have been a
ponderous lament a witty turn by having the singer drunkenly
allude to a story he never tells. Despite his protests that "I've
gotta lotta things to say" and "You simply gotta listen to me, until
it's talked away," he never really tells Joe the "little story you
oughta know." Instead, he wallows in his own mood:

> I'm feelin' so bad,
> I wish you'd make the music dreamy and sad

Nevertheless, Mercer wittily has him *think* he has told his tale:
"That's how it goes," he concludes, then blames his cessation on
his listener—"And Joe, I know you're anxious to close"—and
even apologizes for "bending your ear." Only at the very end, in
an image that confirms his boast that he is a "kind of poet," does
he show his realization that he must—and yet cannot—tell his
story:

> This torch I've found,
> must be drowned
> or it soon might explode

By *not* telling that tale, of course, Mercer achieves an understatement, an "iceberg" effect, as Hemingway termed it, all the more moving than any story that *could* be told.

In "Ac-cent-tchu-ate the Positive" (1944) Mercer and Arlen gave their distinctive twist to yet another form, the spiritual, casting their adaptation in the guise of a sermon. According to one story, they were out for a drive one day in Hollywood, when Arlen began humming a "spiritual-like" melody and Mercer "irrelevantly" tossed up the line "You've got to accentuate the positive."[7] According to another story, Mercer had remembered the line from his southern childhood; still another tale has Mercer taking it from a newspaper clipping about the Harlem revivalist "Father Divine."[8] Whatever the source, the lyric combined the black and southern strains Mercer wove so well.

Like Gershwin in "It Ain't Necessarily So," he uses the clash of slang and sententious diction, setting phrases like "Don't mess" and "latch on" next to highfalutin terms like "affirmative" and "up to the maximum," or producing such back-to-back clashes as "pandemonium li'ble." Rather than patronizing the caricatured preacher, Mercer shares his delight in language that ranges from the sonorous heights to the streets:

> To illustrate my last remark,
> Jonah in the whale, Noah in the ark,
> what did they do just when ev'rything looked so dark?
> "Man," they said, "We better accent-tchu-ate the positive!"

The cadences and syntactical shifts capture a dramatic delivery style that swerves from the pontifical to the chatty, right down to the " 'Man ' they said" that slides, with Arlen's riff, from the release into the final A section.

When Arlen's music moved away from such blues, jazz, and spiritual idioms, however, Mercer's language lost its earthy elegance. Just as Yip Harburg's language soared with Arlen's melody for "Over the Rainbow," Johnny Mercer strained for solemn poetry in the tremulous "My Shining Hour" (1943). Even though they wrote it for Fred Astaire (in a film aptly titled *The Sky's the*

Limit), Mercer's lyric has none of the nonchalant sophistication that was Astaire's hallmark. While some have found it a "romantic hymn" that realizes "tender, absorbing emotion,"[9] Mercer's lyric instead seems to exude the same sentimentality that was rife in the war years. At one point Mercer even alludes to one of the most nostalgic hits of the war, when he envisions

> a glow upon the sky,
> and as time goes by,
> it will never die.

The chorus pushes such imagery to the sky's limit, with "lights of home" and an "angel" who, in a most un-Mercer-like poeticism, is "watching o'er me."

Still, Harold Arlen, when he was in his usual rhythmic and bluesy groove, was Mercer's ideal collaborator; had the partnership endured, it surely would have produced more great standards. When they left Hollywood for the Broadway stage, however, Mercer and Arlen met with failure. According to Martin Gottfried, they failed to realize that, in the wake of *Oklahoma!*, "good songs alone couldn't carry shows."[10] Arlen should have learned that lesson in 1944, when he tried his own Americana piece, *Bloomer Girl*, with Yip Harburg. Even though it starred *Oklahoma!*'s Celeste Holm and was choreographed by Agnes DeMille, *Bloomer Girl* was just "a big, colorful show that made no attempt to depart from the established conventions and formulas of musical comedy."[11]

Like Arlen, Mercer "was not a man of the theater." They each "thought in terms of the song rather than the show," "polishing a song for its own sake rather than the sake of the production." In 1946 they collaborated on *St. Louis Woman*, with a book by Arna Bontemps and Countee Cullen, but the attempt to create black "folk drama" in the tradition of *Porgy and Bess* failed to transcend old-fashioned musicals; the songs, though "classy," were "pop rather than theater songs." While such songs could have made for a successful musical ten years earlier, after *Oklahoma!* they seemed to lack "theatrical spark."[12]

One of them, however, was in the best blues style Arlen and Mercer could muster, and became an independent hit. "Come Rain or Come Shine" took its title from a phrase in "Day In—Day out": "come rain—come shine." Mercer liked the phrase so much he had used a tepid variation of it in 1942 for a movie song he wrote with Jerome Kern, "Dearly Beloved"—"I know that I'll be yours come shower or shine." The phrase got a far better rebirth in 1946, however, when, after Mercer came up with the opening line for Arlen's melody, "I'm gonna love you like nobody's loved you," the composer suggested, "Come hell or high water." "Of course," Mercer said, "Why didn't I think of that . . . Come rain or come shine."

Arlen's music is full of his characteristic repeated notes and octave-drops, and Mercer's lyric is equally characteristic—full of driving, mirror-like phrases:

> Happy together, unhappy together . . .
> We're in or we're out of the money . . .
> Gonna love you . . . gonna be true . . .

As these echoes build, the tone becomes almost threatening, particularly with such harsh avowals of love as "Don't ever bet me" and the reversal of the opening phrase into "You're gonna love me like nobody's loved me." The lyric also has a wide allusive range, from a direct lift from Porter—"it was just one of those things"—to the Depression anthem "We're in the Money."

Just as Ted Koehler shrewdly stretched the closing word of "Stormy Weather" over two of Arlen's notes, Mercer draws his final "shine" over three notes, realizing, as Martin Gottfried points out, "that Arlen's notes were meant to be sung as a blues slide and that individual syllables would have made the song too formal, too racially white." "Exquisite as this lyric is," Gottfried laments, "it simply isn't the sort that contributes to a show. . . . It is a pop lyric."[13] While such a song might have carried a show at the height of the golden age (as "Night and Day" carried *Gay Divorce*), *St. Louis Woman* closed after only 113 performances. The fact that "Come Rain or Come Shine" pledges eternal togetherness makes it all the more sadly ironic that it was the last hit Mercer and Arlen wrote together.

The failure of *St. Louis Woman* was yet one more sign that, for a popular song lyricist like Johnny Mercer, Broadway was becoming a closed shop; as his friend Bob Bach notes, "the 1946 debacle should have warned him of future disappointments."[14] Still, he tried again in 1950, but *Texas Li'l Darlin'* was even less successful, particularly after such integrated musicals as *South Pacific*, *Guys and Dolls*, and *Kiss Me, Kate*. Undaunted, Mercer kept trying throughout the 1950s, with *Top Banana* (1951), *Li'l Abner* (1956), and *Saratoga* (1959)—all flops, even though the last, Bach laments, "had the greatest of ingredients—book by Edna Ferber, costumes and sets by Cecil Beaton, and music by Harold Arlen— and the worst of reviews."

Hollywood, too, increasingly had little to offer lyricists. In the early 1940s it had given Mercer the chance to work not only with Harold Arlen but with other composers, such as Hoagy Carmichael and even the old master himself, Jerome Kern. The unlikely pair of Mercer and Kern even had a hit, "I'm Old Fashioned" (1942), though as the title itself indicates, it was Kern's style, not Mercer's, that dominated; like practically all of Kern's lyricists, Mercer dropped his vernacular guard and swung for such poetic heights as "the starry song that April sings." By 1945, however, Hollywood was giving Mercer the kind of "theme-song" piece work that went back to the earliest days of movie songs. The nonmusical film *Laura*, for example, had a haunting title song that became popular as an instrumental, and the studio, sensing a way to double its money, called Mercer in to put words to it. The assignment was particularly difficult, for not only was David Raksin's melody complex and wide-ranging, but Mercer tried to fit the song to the murder-mystery plot, combining a touch of Southern Gothic—"footsteps that you hear down the hall"—with sentimental nostalgia: "she gave your very first kiss to you . . . but she's only a dream."

Setting lyrics to movie theme songs soon became Mercer's

major outlet for his talents. He still had occasional opportunities to write for film musicals: "On the Atchison, Topeka, and the Santa Fe" was a big production number for Judy Garland in *The Harvey Girls*, and *Here Comes the Groom* reunited Mercer and Carmichael in the salty regionalism of "In the Cool, Cool, Cool of the Evening" (1951). But, except for MGM, which would hold out into the early 1950s, the big studio empires on which film musicals depended were crumbling. Lavish efforts of 1948, such as *Summer Holiday* and *The Pirate* were box-office flops, and Hollywood began to rely on the safer practice of filming established Broadway musicals. In 1954 Johnny Mercer wrote the lyrics for *Seven Brides for Seven Brothers*, generally regarded as the last of the great, original Hollywood musicals, yet one which generated no successful popular songs. What few original musicals Hollywood did turn out were low-budget fare, geared to a teenage audience, though Mercer still tried to hang in there, writing "Bernadine" (1958) for a Pat Boone film.

An ironic sign of the demise of film musicals was *Daddy Long Legs*, where an aging Fred Astaire was paired with Leslie Caron. Faced with an impasse—how to create romance between an old and a young dancer—the studios called in Mercer to write a song to resolve the problem. Mercer cured the situation with "Something's Gotta Give" (1954), a song that went back to the nonchalant sophistication of "I'm Building Up to an Awful Let-Down." *Daddy Long Legs* was one of Astaire's last pictures; "Something's Gotta Give," the last song he made popular. For Mercer, too, it was the last of his film songs in Astaire's—and the golden age's—urbane style. His others were simple movie theme songs, usually setting the lush soarings of Henry Mancini with such nostalgic pieties as "Moon River" (1961) and "Days of Wine and Roses" (1962).

If Broadway and Hollywood were changing, Tin Pan Alley was disappearing. The demise of New York music publishers as the dominant force in popular song had begun in the 1930s, when, as we have seen, Hollywood studios bought out the major publishing houses. It was hastened in the 1940s, when record sales finally displaced sheet music as the major money-maker in popular music. Radio corporations like RCA and film studios like Columbia quickly created their own recording companies, and independent companies began to spring up as well—ironically, one of the first and most successful was Capitol Records, founded in 1941, by Mercer himself, along with Buddy DeSylva and Glen Wallichs.

Record companies soon began aiming their wares at a younger audience—the bobby-soxers who swooned for Sinatra in ballrooms and then around the jukeboxes springing up across the country. Mercer himself contributed to his company's profits with such jukebox fare as "G.I. Jive" (1944), and he also sang saccharine wares like "Dream" (1944) over his weekly radio program. In the end it was the very commercialism that spawned Tin Pan Alley that buried it. In 1941, ASCAP, the American Society of Composers, Authors and Publishers, an organization created in 1914 by Victor Herbert and other early songwriters to protect their profits, demanded that radio stations pay twice as much in royalties for the right to play its music. Rather than pay the double price, radio stations organized their own performing rights society: Broadcast Music, Incorporated. Linking up with record companies in Chicago, companies that for years had been combing the South—Mercer's South!—for material, BMI acquired the songs of writers who up to then "because of ASCAP had been excluded from the charmed circle—the hillbillies and the blacks."

When their license to play ASCAP material ran out at midnight, January 1941, radio stations began playing songs so old they were in public domain—Stephen Foster, for example, had a revival—and new songs by writers from Mercer's backyard:

> those who benefited immediately were the hillbillies, who by now had their own well-established network of local country radio shows, and the black musicians, who for years had been recorded for little or no financial reward. The ASCAP monopoly was broken and the absolute domination of Tin Pan Alley came to an end.[15]

Although the boycott was lifted and ASCAP signed a new agreement with radio, BMI had opened up the industry to black and country composers, and within a few years such old Alley April-shower banalities as "I believe for every drop of rain that falls a flower grows" would be rudely displaced by warnings to "lay offa my blue suede shoes."

Within a few years, the influx of black and country music had completely transformed popular music and displaced Tin Pan Alley as the musical center of the universe. With the Alley went its sturdiest pillar—the thirty-two-bar AABA chorus. "After World War II," Gerald Mast notes, "rock and country-western songs deliberately reverted to archaic eight- and sixteen-bar forms, suggesting the aspiration to sincerity, purity, and 'naturalness' by rejecting the urban and urbane song structure of white European immigrants."[16] Thus the strophic, verse-chorus songs of nineteenth-century music that had been abandoned for the Alley "ballad" formula of saying "I love you" in thirty-two bars returned and, with it, a truly narrative ballad with a wider range of subject and theme.

While there were numerous hits in the old Alley style—Mercer himself proved he could still hit the top of the charts as late as 1965 with "Summer Wind"—the shifting audience and market for popular song left Johnny Mercer all the more limited as a lyricist. By the 1950s he was working, like Mitchell Parish, without a collaborator, setting words to instrumentals. In some cases these were European songs, such as "Autumn Leaves" (1950) and "The Glow Worm" (1952), which had already proven their popularity abroad. Such lyrics took him far from his regional roots, like his setting for M. Philippe-Gerard's chanson "When the World Was Young" (1951):

> Summers at Bordeaux,
> rowing the bateau

Still, the nostalgic narrative of a sophisticated "boulevardier, the toast of Paris," who recalls with envy his rural youth—"ah, the apple trees, blossoms in the breeze"—could easily describe Mercer himself.

Perhaps no image better captures Mercer's isolation in these twilight years than the story he himself told about writing the lyric to "Midnight Sun" (1954), which had originally been an instrumental by Lionel Hampton and Sonny Burke. Driving along the freeway from Palm Springs to Hollywood, he heard the song on his car radio and set words to it as he drove. The West Coast setting, the car, the lyricist reduced to working with his composers via radio—how far removed from the bustling music publishing houses that once lined Tin Pan Alley. In "Midnight Sun," the oldest clichés of the Alley, appropriately it would seem, are pushed to baroque extremes: lips

> like a red and ruby chalice,

clouds

> like an alabaster palace,

and every star

> its own aurora borealis.

It's as if the lyric itself is a midnight sun, a last blaze of an Alley style extinguishing itself along with the Broadway stage and Hollywood studios its songs once had fueled.

Still, as late as 1958, Johnny Mercer could write one last lyric with the vernacular elegance—and his own driving, slangy phrasing—of the golden age, a lyric that could well serve as the era's epitaph. The fact that "Satin Doll" was set to an Ellington instrumental would make anyone wish the two had been permanent collaborators. In Ellington's music, as in Arlen's, Mercer found an earthy sophistication that perfectly suited his lyrical style, his distinctive adaptation of the heritage of Hart, Gershwin, and Porter. In "Satin Doll" Mercer makes that style his subject: musically and lyrically, "Satin Doll" is as elegant and brash as the lady it celebrates:

> Cigarette holder
> which wigs me
> over her shoulder
> she digs me

Mercer pushes the blend of slang and sophistication to its limit:

> telephone numbers
> well, you know
> doin' my rhumbas
> with uno

Then, with a final, flippant flourish Mercer deftly rhymes the simplest of words:

> speaks Lat-
> in th*at*
> *sat-*
> in doll

That Latin-speaking, satin-attired epitome of slangy urbanity might well be the muse of all the poets of Tin Pan Alley, glimpsed here in her last incarnation. If so, she was a muse who, for Johnny Mercer, proved a "worrisome" thing indeed, leaving him alone, in the twilight of the golden age, to sing the blues in the night.

Notes

Chapter 1. Blah, Blah, Blah, Blah Love: Alley Standards

1. Susanne K. Langer, *Problems of Art* (New York: Charles Scribner's Sons, 1957), 84.
2. Anthony Burgess, *This Man and Music* (London: Hutchinson, 1982), 105–6.
3. Ira Gershwin, *Lyrics on Several Occasions* (New York: Alfred A. Knopf, 1959), ix.
4. E. Y. Harburg, "From the Lower East Side to 'Over the Rainbow' " *Creators and Disturbers*, eds. Bernard Rosenberg and Ernest Goldstein (New York: Columbia Univ. Press, 1982), 140–41.
5. Carolyn Wells, *A Vers de Société Anthology* (New York: Charles Scribner's Sons, 1907), xix–xxv.
6. John Updike, "Foreword" to *The Complete Lyrics of Cole Porter,* ed. Robert Kimball (New York: Vintage Books, 1984), xiv.
7. Isaac Goldberg, *Tin Pan Alley* (New York: John Day Company, 1930), 230.
8. Gerald Mast, *Can't Help Singin': The American Musical on Stage and Screen* (Woodstock, N.Y.: Overlook Press, 1987), 191.
9. Cleanth Brooks, *The Well-Wrought Urn: Studies in the Structure of Poetry* (London: Dennis Dobson, 1947), 12.
10. Gershwin, *Lyrics on Several Occasions,* 42.
11. Sheila Davis, *The Craft of Lyric Writing* (Cincinnati, Ohio: Writers Digest Books, 1985), 194.
12. Edward Pessen, "The Great Songwriters of Tin Pan Alley's Golden Age," *American Music* 3 (Summer 1985), 192–93.
13. James J. Wilhelm, *Seven Troubadours* (University Park: Pennsylvania State Univ. Press, 1970), 13.

14. In *Thou Swell, Thou Witty,* ed. Dorothy Hart (New York: Harper and Row, 1976), 18.

15. Charles Hamm, *Yesterdays: Popular Song in America* (New York: W. W. Norton, 1979), 377–78.

Chapter 2. After the Ball: Early Alley

1. Warren Craig, *Sweet and Lowdown: America's Popular Songwriters* (Metuchen, N.J.: Scarecrow Press, 1978), 15.

2. David Ewen, *The Life and Death of Tin Pan Alley: The Golden Age of American Popular Music* (New York: Funk and Wagnalls, 1964), 5.

3. Kenneth Kanter, *The Jews on Tin Pan Alley* (New York: Ktav Publishing, 1982), 17.

4. Hamm, *Yesterdays,* 286.

5. Sigmund Spaeth, *A History of Popular Music in America* (New York: Random House, 1948), 259.

6. Mark W. Booth, *The Experience of Songs* (New Haven: Yale Univ. Press, 1981), 163–64.

7. Charles K. Harris, *After the Ball: Forty Years of Melody* (New York: Frank Maurice, 1926), 19.

8. Hamm, *Yesterdays,* 292.

9. Goldberg, *Tin Pan Alley,* 155–56.

10. Eugene D. Levy, *James Weldon Johnson: Black Leader, Black Voice* (Chicago: Univ. of Chicago Press, 1973), 87–88.

11. Spaeth, *A History of Popular Music in America,* 338.

12. Max Morath, "Introduction," *Favorite Songs of the Nineties,* ed. Robert A. Fremont (New York: Dover, 1973), ix.

13. Leonard Bernstein, *The Joy of Music* (New York: Simon and Schuster, 1959), 166–67.

14. Alec Wilder, *American Popular Song: The Great Innovators, 1900–1950* (New York: Oxford Univ. Press, 1972), 14.

15. *Ibid.,* 20.

16. Arnold Shaw, *The Jazz Age* (New York: Oxford Univ. Press, 1987), 77–78.

17. Mast, *Can't Help Singin',* 24.

18. Martin Gottfried, *Broadway Musicals* (New York: Abradale Press, 1984), 166.

19. David Ewen, *Complete Book of the American Musical Theater* (New York: Holt, Rinehart and Winston, 1958), 167.

20. P. G. Wodehouse, *Author! Author!* (New York: Simon and Schuster, 1962), 15.

21. Tony Palmer, *All You Need Is Love: The Story of Popular Music* (New York: Grossman, 1976), 123.

22. Bernstein, *The Joy of Music*, 164.

23. Hamm, *Yesterdays*, 339.

Chapter 3. Irving Berlin

1. According to Tony Hill's unpublished "Chart History of Songs by Irving Berlin," the first singer to have a hit with "How Deep Is the Ocean" was Ethel Merman. I am indebted to Tony Hill for sharing his extensive research on Berlin with me.

2. Wilder, *American Popular Song*, 99.

3. Michael Freedland, *Irving Berlin* (New York: Stein & Day, 1974), 69.

4. Mast, *Can't Help Singin'*, 45.

5. Lewis Erenberg, *Steppin' Out: New York Nightlife and the Transformation of American Culture, 1890–1930* (Westport, Conn.: Greenwood Press, 1981), 242.

6. Wilder, *American Popular Song*, 104.

7. Timothy Scheurer, "Irving Berlin," paper delivered at the American Popular Culture Conference, Atlanta, April 6, 1986.

8. Wilder, *American Popular Song*, 109.

9. Freedland, *Irving Berlin*, 129.

10. Gottfried, *Broadway Musicals*, 240.

11. Mast, *Can't Help Singin'*, 41.

Chapter 4. Lyricists of the 1920s

1. Gottfried, *Broadway Musicals*, 235.

2. Wilder, *American Popular Song*, 453.

3. *Ibid.*, 296.

4. Craig, *Sweet and Lowdown*, 293.

5. Wilder, *American Popular Song*, 404.

6. Craig, *Sweet and Lowdown*, 293.

7. As was customary from 1934 through 1937, the award itself was regarded as a music department achievement and was presented to the head of the studio music department—in this case Louis Silvers, head of the Columbia Studio Music Department. I am indebted to

Mr. Donald Kahn for calling my attention to this fact, as well as for his help with other historical information regarding his father's career.

8. Ewen, *All the Years of American Popular Music: A Comprehensive History* (Englewood Cliffs, N.J.: Prentice-Hall, 1978), 338.
9. Brooks, *The Well-Wrought Urn,* 12.
10. Spaeth, *A History of Popular Music in America,* 474.
11. Ethan Mordden, *The Hollywood Musical* (New York: St. Martin's, 1981), 78.

Chapter 5. Lorenz Hart

1. Samuel Marx and Jan Clayton, *Rodgers and Hart* (New York: G. P. Putnam's Sons, 1976), 38.
2. Oscar Hammerstein, *Lyrics* (New York: Simon and Schuster, 1949), 20.
3. *Thou Swell,* 36.
4. Wilder, *American Popular Song,* 169.
5. Richard Rodgers, *Musical Stages* (New York: Random House, 1975), 201.
6. *Thou Swell,* 35.
7. *Ibid.,* 41.
8. Ewen, *All the Years of American Popular Music,* 346.
9. Burgess, *This Man and Music,* 108.
10. Marx and Clayton, *Rodgers and Hart,* 84.
11. Rodgers, *Musical Stages,* 73.
12. Wilder, *American Popular Song,* 175.
13. Lehman Engel, *Their Words Are Music* (New York: Crown, 1975), 39.
14. *Thou Swell,* 58.
15. Rodgers, *Musical Stages,* 115.
16. Burgess, *This Man and Music,* 109.
17. Rodgers, *Musical Stages,* 164.
18. Marx and Clayton, *Rodgers and Hart,* 211.

Chapter 6. Ira Gershwin

1. Edward Jablonski and Lawrence D. Stewart, *The Gershwin Years* (Garden City, N.Y.: Doubleday, 1958; rev. ed. 1973), 41.
2. *American Popular Song: Six Decades of Songwriters and Singers,* eds.

James R. Morris, J. R. Taylor, and Dwight Blocker Bowers (Washington: Smithsonian Institution Press, 1984), 38–39.

3. In Robert Kimball and Alfred Simon, *The Gershwins* (New York: Atheneum, 1973), 46.
4. Mast *Can't Help Singin'*, 72.
5. *Thou Swell*, 53.
6. Gershwin, *Lyrics on Several Occasions*, 359–60.
7. Mast, *Can't Help Singin'*, 85.
8. Gershwin, *Lyrics on Several Occasions*, 48.
9. Jablonski and Stewart, *Gershwin Years*, 335.
10. Gershwin, *Lyrics on Several Occasions*, 275.

Chapter 7. Cole Porter

1. *Cole*, ed. Robert Kimball (New York: Holt, Rinehart and Winston, 1971), 72.
2. Charles Schwartz, *Cole Porter* (New York: The Dial Press, 1977), 224.
3. Ewen, *All the Years of American Popular Music*, 372.
4. Rodgers, *Musical Stages*, 88.
5. Schwartz, *Cole Porter*, 55.
6. *Ibid.*, 70.
7. Wilder, *American Popular Song*, 226.
8. *Ibid.*, 227.
9. Gottfried, *Broadway Musicals*, 56.
10. Wilder, *American Popular Song*, 235.
11. *Ibid.*, 240.
12. Schwartz, *Cole Porter*, 155.
13. Wilder, *American Popular Song*, 244.
14. *Cole*, 149.
15. Schwartz, *Cole Porter*, 178.
16. *Cole*, xvi.
17. Schwartz, *Cole Porter*, 207.
18. Mast, *Can't Help Singin'*, 182.
19. Gottfried, *Broadway Musicals*, 215.
20. *A Midsummer Night's Dream*, II, i, 204–6. I am indebted to my colleague, Norman Fruman, for pointing out this allusion to me.
21. Schwartz, *Cole Porter*, 256.

Chapter 8. Oscar Hammerstein

1. Hammerstein, *Lyrics* 20.
2. Ewen, *Complete Book of the American Musical Theater,* 91.
3. Hammerstein, *Lyrics,* 41.
4. Gottfried, *Broadway Musicals,* 166–67.
5. Hammerstein, *Lyrics,* 19–20.
6. The original line, "Niggers all work," was changed over the years to "colored folks work," "lots of folk work," and "here we all work."
7. Hammerstein, *Lyrics,* 23.
8. *Ibid.,* 8.
9. Gottfried, *Broadway Musicals,* 175.
10. Hammerstein, *Lyrics,* 9.
11. In *The Popular Arts: A Critical Reader,* eds. Irving Deer and Harriet A. Deer (New York: Charles Scribner's Sons, 1967), 199.
12. Gottfried, *Broadway Musicals,* 63–64.
13. Mast, *Can't Help Singin',* 203.

Chapter 9. Howard Dietz and Yip Harburg

1. Howard Dietz, *Dancing in the Dark* (New York: Quadrangle, 1974), 120.
2. Gottfried, *Broadway Musicals,* 62.
3. Palmer, *All You Need is Love,* 123.
4. Dietz, *Dancing in the Dark,* 124.
5. Engel, *Their Words Are Music,* 63.
6. Wilder, *American Popular Song,* 315.
7. Gottfried, *Broadway Musicals,* 63.
8. Ewen, *All the Years of American Popular Music,* 400.
9. Harburg, "From the Lower East Side to 'Over the Rainbow,' " *Creators and Disturbers,* 145.
10. Max Wilk, *They're Playing Our Song* (New York: Atheneum, 1973), 223.
11. Harburg, "From the Lower East Side to 'Over the Rainbow,' " *Creators and Disturbers,* 146.
12. Wilk, *They're Playing Our Song,* 225.
13. *Ibid.,* 224.
14. Edward Jablonski, *Harold Arlen: Happy With the Blues* (Garden City, N.Y.: Doubleday, 1961), 120.

15. Wilk, *They're Playing Our Song*, 229.
16. Wilder, *American Popular Song*, 338.

Chapter 10. Dorothy Fields and Leo Robin

1. Henry Kane, *How to Write a Song* (New York: Macmillan, 1962), 174.
2. Ewen, *All the Years of American Popular Music*, 343.
3. Jack Burton, *The Blue Book of Tin Pan Alley* (Watkins Glen, N.Y.: Century House, 1951), 323.
4. Wilder, *American Popular Song*, 407.
5. Mast, *Can't Help Singin'*, 58.
6. Wilk, *They're Playing Our Song*, 47–48.
7. *Ibid.*, 98.
8. Mast, *Can't Help Singin'*, 108.
9. Wilk, *They're Playing Our Song*, 103.
10. *Ibid.*, 106.
11. *Ibid.*, 105.
12. Theodore Taylor, *Jule: The Story of Composer Jule Styne* (New York: Random House, 1979), 132.
13. Gottfried, *Broadway Musicals*, 261.

Chapter 11. Hollywood Lyricists

1. Mordden, *The Hollywood Musical*, 78.
2. Ewen, *All the Years of American Popular Music*, 275.
3. Goldberg, *Tin Pan Alley*, 303.
4. Mordden, *The Hollywood Musical*, 75.
5. Mast, *Can't Help Singin'*, 94.
6. Quoted in Mordden, *The Hollywood Musical*, 14.
7. Quoted in *American Popular Songs: From the Revolutionary War to the Present*, ed. David Ewen (New York: Random House, 1966), 388.
8. Wilder, *American Popular Song*, 395.
9. Mordden, *The Hollywood Musical*, 44.
10. Mast, *Can't Help Singin'*, 124.

Chapter 12. Swingy Harlem Tunes: Jazz Lyricists

1. Burton, *The Blue Book of Tin Pan Alley*, 412.
2. Wilk, *They're Playing Our Song*, 152–53.

3. Burton, *The Blue Book of Tin Pan Alley*, 411.
4. Wilder, *American Popular Song*, 262.
5. Burton, *The Blue Book of Tin Pan Alley*, 412.
6. *American Popular Song: Six Decades of Songwriters and Singers*, 58.
7. Wilder, *American Popular Song*, 501.
8. Quoted in *American Popular Songs From the Revolutionary War to the Present*, 366.
9. Wilder, *American Popular Song*, 375.
10. *Ibid.*, 413.
11. *Ibid.*, 414.

Chapter 13. Johnny Mercer

1. Gottfried, *Broadway Musicals*, 254.
2. *Our Huckleberry Friend*, eds. Bob Bach and Ginger Mercer (Secaucus, N.J.: Lyle Stuart, 1974) 50.
3. Wilder, *American Popular Song*, 424.
4. *Ibid.*, 436.
5. *Our Huckleberry Friend*, 98.
6. Wilder, *American Popular Song*, 273.
7. Quoted in *American Popular Songs from the Revolutionary War to the Present*, 4.
8. *Our Huckleberry Friend*, 128.
9. *American Popular Song: Six Decades of Songwriters and Singers*, 106.
10. Gottfried, *Broadway Musicals*, 253.
11. Ewen, *Complete Book of the American Musical Theater*, 12.
12. Gottfried, *Broadway Musicals*, 253–54.
13. *Ibid.*, 66.
14. *Our Huckleberry Friend*, 91.
15. Palmer, *All You Need Is Love*, 112.
16. Mast, *Can't Help Singin'*, 27.

Grateful acknowledgment is made to the following
for kind permission to quote song lyrics in copyright:

291

JoRo Music Corporation

CRAZY RHYTHM (Words by Irving Caesar; Music by Joseph Meyer and Roger Wolfe Kahn). Copyright JoRo Music Corp. Warner Brothers Music, Inc. Reprinted with permission.

IF YOU KNEW SUSIE (Words and Music by Joseph Meyer and B.G. DeSylva). Copyright JoRo Music Corp. Shapiro Bernstein and Co., Inc. Stephen Ballentine Music Co. Reprinted with permission.

Gus Kahn Music Coporation

AIN'T WE GOT FUN (Gus Kahn-Raymond Egan-Richard Whiting). Used with the permission of the copyright owner.

CAROLINA IN THE MORNING (Gus Kahn-Walter Donaldson). Used with the permission of the copyright owner.

I'M THRU WITH LOVE (Gus Kahn-Matt Malneck-Fud Livingston). Used with the permission of the copyright owner

IT HAD TO BE YOU (Gus Kahn-Isham Jones). Used with the permission of the copyright owner.

LOVE ME OR LEAVE ME (Gus Kahn-Walter Donaldson). Used with the permission of the copyright owner.

MAKIN' WHOOPEE (Gus Kahn-Walter Donaldson). Used with the permission of the copyright owner.

THE ONE I LOVE BELONGS TO SOMEBODY ELSE (Gus Kahn-Isham Jones). Used with the permission of the copyright owner.

YES SIR, THAT'S MY BABY (Gus Kahn-Walter Donaldson). Used with the permission of the copyright owner.

MPL Communications

THERE WILL NEVER BE ANOTHER YOU (Harry Warren and Mack Gordon). © 1942 Twentieth Century Music Corporation; © renewed 1970 Twentieth Century Music Corporation. All rights throughout the world controlled by Morley Music Co. International copyright secured. All rights reserved. Used by permission.

AC-CENT-TCHU-ATE THE POSITIVE (Johnny Mercer and Harold Arlen). © 1944 Harwin Music Co.; © renewed 1972 Harwin Music Co. International copyright secured. All rights reserved. Used by permission.

MY SHINING HOUR (Johnny Mercer and Harold Arlen). © 1943 Harwin Music Co.; © renewed 1971 Harwin Music Co. International copyright secured. All rights reserved. Used by permission.

ONE FOR MY BABY (AND ONE MORE FOR THE ROAD) (Johnny Mercer and Harold Arlen). © 1943 Harwin Music Co.; © renewed 1971 Harwin Music Co. International copyright secured. All rights reserved. Used by permission.

HOORAY FOR LOVE (Leo Robin and Harold Arlen). © 1948 Harwin Music Co.; © renewed 1976 Harwin Music Co. International copyright secured. All rights reserved. Used by permission.

Index

ASCAP (American Society of Composers, Authors and Publishers), 42, 279–80
"Ac-cent-tchu-ate the Positive," 274
Adair, Tom, 44–45, 79
Adams, Franklin Pierce (F.P.A.), 7, 8, 196, 197, 203, 215
Adios, Argentina, 177
"After the Ball," 22–25, 41, 44, 47, 183
"After You've Gone," 37
"Ah! Sweet Mystery of Life," 31, 51
"Ain't Misbehavin'," 11, 87
"Ain't We Got Fun?," 76–77
Akst, Harry, 244
Alda, Frances, 58
"Alexander, Don't You Love Your Baby No More," 49
"Alexander's Ragtime Band," 49–50, 51, 128
Algonquin Round Table," 18, 99
"All Alone," 55, 56, 58, 77, 86
"All by Myself," 37, 47, 55–56, 59, 86
"All Coons Look Alike to Me," 26
"All I Do Is Dream of You," 235
"All of You," 179
"All the Things You Are," 9, 188, 190
"All Through the Night," 171, 172
"Alone Together," 201
"Always," 55
"Am I Blue?," 36
"American Language, The, 11
Americana, 196, 203
"And the Angels Sing," 268
Anderson, John Murray, 159
Anderson, Maxwell, 200
"Angel Eyes," 254, 273

Annie Get Your Gun, 47, 68–70, 222
"Any Bonds Today?," 67
"Any Old Place with You," 97–98, 100
Anything Goes, 93, 165, 171
"Anything Goes," 154, 155, 168
"Anything You Can Do," 68
"April in Paris," 205–7
"April Showers," 88
"Araby," 53
Arlen, Harold, 3, 37, 63, 94, 151–52, 208–11, 229, 244, 245–50, 254, 263, 264, 270–77, 281
Armory Show of 1913, 10
Armstrong, Louis, 246
As Thousands Cheer, 63
"As Time Goes By," 67, 241
Astaire, Adele, 128
Astaire, Fred, 18, 47, 64–66, 82–83, 144, 145–48, 149, 202, 219–22, 237, 265, 274–75, 278
"At a Georgia Camp Meeting, 27
"At Long Last Love," 175–76
Atkinson, Brooks, 123
"Autumn in New York," 207
"Autumn Leaves," 280

BMI (Broadcast Music Incorporated), 279–80
"Bab Ballads," 6
Babes in Arms, 117, 119–21
Bach, Bob, 266, 277
"Bad in Every Man, The," 116
Balanchine, George, 5
Baline, Leah, 48
Baline, Moses, 47
Band Wagon, The, 196, 200, 203
Baum, Frank, 209

Baxter, Phil, 240
Bayes, Nora, 32
Beaton, Cecil, 277
"Because (you come to me)," 32
Beggar's Opera, The, 13
"Begin the Beguine," 171–72
Beiderbecke, Bix, 246, 256
Benchley, Robert, 18, 99, 197
Benny, Jack, 226
Benton, Thomas Hart, 125, 189
Berkeley, Busby, 116, 238
Berlin, Irving, 22, 43, 46–71, 77, 81,
    86, 87, 92, 97, 101, 128, 129, 144,
    154–55, 165, 168, 173, 174, 177,
    200, 215, 216, 222, 231, 237, 248–
    49, 252, 264
"Bernadine," 278
Bernstein, Leonard, 31–32, 41
Bernstein, Louis, 20
"Bess, You Is My Woman Now," 142
Between the Devil, 196, 206
"Between the Devil and the Deep
    Blue Sea," 247
"Bewitched," 97, 123–24, 211
"Beyond the Blue Horizon," 225
"Bidin' My Time," 138–39
Big Broadcast of 1938, 227
"Big Spender," 222–23
Bigard, Albany, 259
"Bill," 40, 112, 184
"Bill Bailey, Won't You Please Come
    Home?," 29, 33
"Bird in a Gilded Cage, A," 28, 257
Birth of a Nation, The, 232
"Birth of the Blues, The," 36, 89
Bishop, Morris, 8
Blackbirds of 1928, 216
Bloom, Rube, 268–70
Bloomer Girl, 210, 275
"Blue Hawaii," 229
"Blue Moon," 116, 177
"Blue Room, The," 96, 108, 112–13
"Blue Skies," 47, 59–60, 233
"Blues in the Night," 151, 263, 270–72
Bolton, Guy, 105
Bontemps, Arna, 275
Boone, Pat, 278
Booth, Mark, 23
Bordoni, Irene, 160
Boys from Syracuse, The, 117, 121
"Break the News to Mother," 25

Brent, Earl, 254
Brice, Fanny, 34, 233
Broadway Melody of 1929, 234, 235
Broadway Melody of 1938, 34
Broadway Melody of 1940, 174
Brooks, Cleanth, 10, 89–90
Brooks, Shelton, 33
"Brother, Can You Spare a Dime?,"
    203–5, 210
Brown, Lew, 87–94, 106, 233
Brown, Nacio Herb, 235–36, 237, 240
Browning, Elizabeth Barrett, 60
"Brush Up Your Shakespeare," 178
Burgess, Anthony, 4–5, 7, 9, 104, 108,
    112
Burke, Johnny, 253
Burke, Joseph, 237–38
Burke, Sonny, 281
Burns, Robert, 6
Burton, Jack, 245
"But in the Morning, No," 176
"But Not for Me," 12, 137–38
Butterfield, Billy, 253
"Button Up Your Overcoat," 91–92
By Jupiter, 124–25
"By Myself," 202
By the Beautiful Sea, 222
"By the River Sainte Marie," 237
"Bye, Bye Blackbird," 85–86
Bynner, Witter, 8
Byrd, Richard, 127

Caesar, Irving, 2, 19, 72–75, 76, 83, 128
Call Me Madam, 70
Calloway, Cab, 245, 248, 258
Campion, Thomas, 13
"Can I Forget You?," 188, 190
Can-Can, 178–79
"Can't Help Lovin' Dat Man," 184
Cantor, Eddie, 75, 88
Capitol Records, 279
"Caravan," 259
"Carioca, The," 82
Carmen, 189
Carmen Jones, 189
Carmichael, Hoagy, 4, 255–56, 264,
    265, 277, 278
"Carolina in the Morning," 77
Carroll, Harry, 43
Carroll, Lewis, 6
Carter, Desmond, 199–200

Casablanca, 67, 241
Castles, The (Irene and Vernon), 50
Catlett, Walter, 134
"Change Partners," 66
Channing, Carol, 230
Charig, Phil, 134
"Charley, My Boy," 77
"Charmaine," 233
"Chattanooga Choo-Choo," 241
"Cheek to Cheek," 47, 64–65
Chevalier, Maurice, 114–15, 224–25
Chopin, Frederic, 43
Clayton, Jan, 95n, 105n, 122n
Cohan, George M., 29–31, 37, 52
Cole, Bob, 27–28
Coleman, Cy, 222
"College on Broadway, A," 98
Columbia Pictures, 279
Columbia University *Varsity Show*,
   98–99
"Come Rain or Come Shine," 276–77
Connecticut Yankee, A, 106–9, 125
"Conning Tower," 7, 8, 196, 197, 203,
   215
"Convict and the Bird, The," 14
Cook, Will Marion, 246
Coots, J. Fred, 215, 252, 254
Copland, Aaron, 125
Cotton Club, 83, 214, 216, 222, 245,
   249–50, 256, 257, 258, 270
Coward, Noel, 157
Craig, Warren, 20n, 76, 82
Crosby, Bing, 47, 225, 226, 236
Crump, Mayor Edward H., 35
Cubism, 10–11, 62, 129, 166
Cullen, Countee, 275
cummings, e. e., 10, 11, 166

Dadaism, 10–11
Daddy Long Legs, 278
Dali, Salvador, 92
Damsel in Distress, A, 144, 147, 148
Dance of Life, 224
Dancing in the Dark, 195
"Dancing in the Dark," 200–201
"Dancing in the Moonlight," 83
"Dancing on the Ceiling," 114
"Dancing with Tears in My Eyes," 238
"Dar's No Coon Warm Enough for
   Me," 29

Davis, Sheila, 14n
Day, Doris, 14
Day at the Circus, A, 208
"Day by Day," 11
"Day In—Day Out," 11, 268–70, 271,
   272, 276
"Days of Wine and Roses," 264, 278
"Dearly Beloved," 276
"Deed I Do," 83
"Deep Purple," 257
Del Rio, Dolores, 233
DeLange, Eddie, 258
Delicious, 144
DeMille, Agnes, 275
Dennis, Matt, 254
DeRose, Peter, 257
DeSylva, B.G. ("Buddy"), 73, 87–94,
   106, 128, 160, 233, 279
"Diamonds Are a Girl's Best Friend,"
   230
Dickinson, Emily, 255
"Did You Ever See a Dream Walk-
   ing?," 241
Dietz, Howard, 7, 8, 94, 96, 103, 146,
   195–203, 207, 215, 219
"Diga Diga Doo," 214
"Dinah," 244
Dinner at Eight, 218
Dixon, Mort, 85–86
"Do It Again," 160
"Do Nothin' Till You Hear from Me,"
   261
Donaldson, Walter, 77, 80–82, 87
Donnelly, Dorothy, 215
"Don't Blame Me," 218
"Don't Fence Me In," 177
"Don't Get Around Much Anymore,"
   11, 260–61
Doraldina, 53
"Dorando," 46, 47, 49
Dowland, John, 13
Dowson, Ernest, 6
"Dream," 279
"Dream a Little Dream of Me," 83
Dreyfus, Max, 158, 180
"Drink to Me Only with Thine Eyes,"
   6
Du Barry Was a Lady, 176
Dubin, Al, 237–40, 241, 243
Duke, Vernon, 142, 205–7
Dunne, Irene, 188

Durante, Jimmy, 73
Dynamite, 231

"Eadie Was a Lady," 93
"Easter Parade," 47, 63, 68
"Easy to Love," 173–74
Eddy, Nelson, 172–73, 209
Edmund's Cellar, 244–45
Egan, Raymond, 76
Elder, Ruth, 215
Eliot, T. S., 10, 58–59
Eliscu, Edward, 82, 85, 223
Ellington, Duke, 3, 216, 243, 245,
   257–62, 264, 281
Engel, Lehman, 108–9, 198
Erenberg, Lewis, 58
Etting, Ruth, 34, 83, 112–14
"Ever and Ever Yours," 157
Every Night at Eight, 218
"Everybody Step," 61
"Everything Happens to Me," 11, 45,
   79
"Ev'rything I Love," 176
Ewen, David, 20n, 39, 89n, 99n,
   232n, 275n

"Faded Summer Love, A," 240
Fagan, Barney, 27
Fantaisie Impromptu, 43
"Fascinating Rhythm," 75, 128–30
Feist, Leo, 20, 21
Ferber, Edna, 183, 277
Fields, Dorothy, 3, 7, 68, 83, 203,
   213–23, 226, 229, 237, 245, 256
Fields, Herbert, 68, 97, 105, 215, 222
Fields, Lew, 97–98, 214
Fifty Million Frenchmen, 63, 163
"Fine Romance, A," 220–21
Finian's Rainbow, 210–12
Fiorito, Ted, 77
Firebrand of Florence, The, 149
"Fish, The," 10, 105
Fitzgerald, F. Scott, 77, 137
"Five Foot Two," 84
Fletcher, Robert, 177
Flying Down to Rio, 82
"Foggy Day, A," 144, 147
Follow the Boys, 241
Fonda, Henry, 267
"Fools Rush In," 270
"For Me and My Gal," 250

Forty-Second Street, 116, 238
Foster, Stephen, 6, 20, 279
"Francis, Arthur" (*see* Ira Gershwin)
Free for All, 181
Freed, Arthur, 209, 235–36, 237, 240
Freedland, Michael, 55n, 67n
"Friendship," 176
Friml, Rudolf, 31, 37, 182
"From Here to Shanghai," 53
"From This Moment On," 11, 179
Fruman, Norman, 178n
Fulton, Robert, 126
"Fun To Be Fooled," 207
Funny Girl, 230

"G.I. Jive," 279
Gable, Clark, 34
"Gal That Got Away, The," 111
Gandhi, Mahatma, 11
Garland, Judy, 34, 278
Garrick Gaieties, 16, 99, 103–5, 182
Garrick Theatre, 99
Gaskill, Clarence, 258
Gay, John, 13
Gay Divorce, 164, 276
Gensler, Lewis E., 225
Gentlemen Prefer Blondes, 229–30
George, Don, 3, 261–62
Gerber, Alex, 98
Gershwin, Arthur, 128
Gershwin, Frances, 128
Gershwin, George, 3, 12, 16, 18, 21,
   40–41, 63, 73, 75, 82, 88, 89, 103,
   117, 124, 126–49, 152, 160, 176,
   181, 189, 213, 237, 246, 248, 250
Gershwin, Ira, 3, 4, 5, 6, 7, 12, 14, 15,
   16, 17, 28, 40, 42, 63, 67, 70, 71,
   73, 75, 82, 88, 93, 94, 97, 103, 110,
   117, 124, 126–52, 153, 154, 160,
   161, 167, 169, 181, 188, 189, 190,
   192, 196, 198, 203, 213, 215, 227,
   231, 237, 266, 270, 274, 281
"Get Happy," 247
Gilbert, W. S., 5, 6, 13, 30, 39–40,
   103, 109, 139, 149, 156, 185, 264
Gill, Brendan, 175
Gillespie, Haven, 252–53, 254
Girl Crazy, 136–39, 144
Girl from Utah, The, 37
"Give My Regards to Broadway," 30
"Glad To Be Unhappy," 118

"Glow Worm," 280
"God Bless America," 47, 67
Goldberg, Isaac, 8n, 26
Goldwyn Follies of 1938, 145, 148
"Good-bye, Little Dream, Good-bye,"
 172
"Goodbye, My True Love," 156–57
Goodman, Benny, 268
"Goody Goody," 268
Gordon, Mack, 237, 240–43
Gorney, Jay, 203
Gorrell, Stu, 255
Gottfried, Martin, 37, 68, 73n, 178,
 183n, 190n, 193, 196n, 203, 230,
 264n, 275, 276
Gray, Glen, 252
Great Day, 246
Great Gatsby, The, 77
Great Magoo, The, 207
Green Grow the Lilacs, 125, 187, 189–90
Greenwich Village Follies of 1924, 159
Grey, Clifford, 224
Griffith, D. W., 232
"Growing Pains," 222
Guiterman, Arthur, 8
Guys and Dolls, 277
Gypsy, 230

Haggart, Bob, 253
Hall, Adelaide, 214
"Hallelujah!," 224
Hamm, Charles, 17, 22, 24, 41–43
Hammerstein, Dorothy (Mrs. Oscar),
 3
Hammerstein, Oscar, 3, 9, 16, 18, 41,
 68, 69, 95, 96, 99, 122, 125, 146,
 181–94, 196, 207, 208, 213, 219,
 263
Hampton, Lionel, 281
Handy, W.C., 35
Happiest Girl in the World, The, 210
"Happiness Is Just a Thing Called
 Joe," 210
"Happy Holiday," 47
Harbach, Otto, 3, 182, 187, 219
Harburg, E.Y. ("Yip"), 3, 6, 7, 8, 94,
 195, 196, 197, 203–12, 215, 244,
 250, 274, 275
Harlow, Jean, 116
Harms, T. B., 158, 233
Harney, Ben, 26

Harris, Charles K., 19, 22–25, 26
Hart, Lorenz, 5, 7, 10, 15, 16, 17, 18,
 28, 42, 63, 67, 70, 71, 73, 93, 94,
 95–125, 126, 129, 130, 132–33,
 140, 145, 153, 154, 160, 161, 163,
 165, 169, 176, 177, 178, 181, 182,
 184, 185, 188, 189, 190, 192, 196,
 197, 198, 199, 200, 204, 205, 206,
 215, 216, 227, 231, 234, 266, 281
Harvey Girls, The, 278
"Heat Wave," 63
Hecht, Ben, 207
"Hello, Central, Give Me Heaven," 25
Hello Daddy, 217
"Hello, Ma Baby," 29
Hemingway, Ernest, 15, 274
Henderson, Fletcher, 246
Henderson, Ray, 85–86, 87–94, 233
Herbert, Victor, 31, 37, 105, 240, 265,
 279
Here Comes the Groom, 278
"Here in My Arms," 106
Here Is My Heart, 225
Herrick, Robert, 6
Herzig, Sig, 16, 18
Heyward, Dorothy, 140
Heyward, DuBose, 140–41
High Society, 179–80
High, Wide and Handsome, 187–88
Higher and Higher, 122
Hill, Tony, 47n
Hirsch, Walter, 83
Hit the Deck, 224
Hitchcock, Raymond, 158
Hitchy-Koo, 1919, 158
Hogan, Ernest, 26, 27
Hold Everything, 89
Holiday, Billie, 252
Hollywood Canteen, 177
Hollywood Party, 116
Holm, Celeste, 275
"Hooray for Hollywood," 268
"Hooray for Love," 229
Hooray for What, 196, 208
Hoover, Herbert, 226
Hope, Bob, 227
Hopper, Hedda, 268
Horne, Lena, 245, 252
"Hot Coon from Memphis, A," 29
"Hot-House Rose," 154
Hot Nocturne, 270–71

"How About Me?," 11, 47, 60
"How Are Things in Glocca Morra?," 210
"How Deep Is the Ocean?," 47, 60
"How Long Has This Been Going On?," 11, 15, 35, 150
"How Do I Love Thee?," 60
Hubbell, Raymond, 128
Huckleberry Finn, Adventures of, 5
Huston, Walter, 200
Hutchinson Family, The Singing, 20

"I Ain't Got Nobody," 37
"I Cain't Say No," 191
"I Can't Begin to Tell You," 243
"I Can't Get Started," 11, 142–44
"I Can't Give You Anything But Love," 31, 216–17, 245
"I Concentrate on You," 174
"I Didn't Know About You," 261
"I Didn't Know What Time It Was," 11, 122
"I Don't Care," 32
"I Don't Know Why I Love You So," 262
"I Don't Stand a Ghost of a Chance," 11
"I Feel a Song Comin' On," 219
"I Found a Million Dollar Baby in a Five and Ten Cent Store," 237
"I Get a Kick Out of You," 169–71, 253
"I Got It Bad and That Ain't Good," 260
"I Got Lost in His Arms," 68
"I Got Plenty o' Nuttin'," 141
"I Got Rhythm," 136, 141
"I Got the Sun in the Morning," 68
"I Gotta Right to Sing the Blues," 247–48
"I Guess I'll Have to Change My Plan," 11, 16, 197–98, 202
"I Guess I'll Have to Telegraph My Baby," 30
"I Had the Craziest Dream," 24
"I Have No Words," 199
"I Let a Song Out of My Heart," 259–60
"I Love a Piano," 51–52
"I Love Paris," 179
"I Love to Lie Awake in Bed," 197

"I Love You," 177
"I Love You Truly," 32, 215
I Married an Angel, 122
"I Only Have Eyes for You," 238–39
"I Should Care," 11
"I Want to Go Back to Michigan Down on the Farm," 53
"I Want Yer, Ma Honey!," 26
I Wish I Were in Love Again," 7, 120–21
"I Won't Dance," 219–20
Ibsen, Henrik, 99
"If I Had a Talking Picture of You," 92–93
"If My Friends Could See Me Now," 223
"If That's Your Idea of a Wonderful Time, Take Me Home," 53
"If There Is Someone Lovelier Than You," 202
"If You Knew Susie," 88
"I'll Build a Stairway to Paradise," 88
"I'll Buy You a Star," 222
"I'll Get By," 241
I'll See You in My Dreams, 75
"I'll See You in My Dreams," 79
"I'll Take You Back to Italy," 53
"I'm Always Chasing Rainbows," 43, 112
"I'm an Old Cowhand," 266
"I'm Beginning to See the Light," 11, 261–63
"I'm Building Up to an Awful Let-Down," 265–66, 278
"I'm Forever Blowing Bubbles," 44
"I'm in Love Again," 159
"I'm in the Mood for Love," 218
"I'm Old-Fashioned," 277
"I'm Shooting High," 250
"I'm Thru with Love," 83
Imagist poetry, 51, 90, 226
"In a Sentimental Mood," 258–59
"In a Station of the Metro," 90
"In Love in Vain," 229
"In the Cool, Cool, Cool of the Evening," 278
"In the Shade of the Old Apple Tree," 76
"In the Still of the Night," 172–73
"Inamorata," 243
"Indian Love Call," 182

"Indian Summer," 240
"Indigo Echoes," 262
Inside U.S.A., 203
International Revue, 217
"Isn't It a Pity?," 140
"Isn't It Romantic?," 114–15
"It Ain't Necessarily So," 141–42, 274
"It Don't Mean a Thing If It Ain't Got
  That Swing," 258
"It Had To Be You," 78–79, 83
"It Never Entered My Mind," 11, 44,
  122
"It Takes an Irishman to Love," 53
"It's All Right With Me," 11, 179
"It's De-lovely," 167–68
"It's Only a Paper Moon," 207–8
"I've a Shooting Box in Scotland," 157
"I've Come to Wive It Wealthily in
  Padua," 178
"I've Got a Crush on You," 135–36
"I've Got a Gal in Kalamazoo," 241
"I've Got the World on a String," 247
"I've Got You Under My Skin," 165,
  174–75

Jablonski, Edward, 127n, 148
Jackson, Arthur, 128
Jacobs-Bond, Carrie, 215
James, Harry, 34
Jazz Singer, The, 47, 92, 233, 234
"Jeepers Creepers," 267–68
Jerome, M.K., 250
"Johnny One-Note," 119
Johnson, J. Rosamond, 27
Johnson, James Weldon, 27
Jolson, Al, 33–34, 47, 88, 92, 141,
  233, 234
Jones, Isham, 78–80
Jubilee, 165
Jumbo, 117–18
"June in January," 226–27
"Just a Girl That Men Forget," 237
"Just A-Wearying for You," 215
"Just Behind the Times," 25
"Just One of Those Things," 11, 169,
  228

Kahn, Donald, 83n
Kahn, Gus, 14, 73, 75–83, 101, 130
Kahn, Roger Wolfe, 75
Kalmar, Bert, 236, 237, 240

Kanter, Kenneth, 21
"Kansas City," 191
Katz, Sam, 172
Kaufman, George S., 55, 126, 139–
  40, 223–24
Kaye, Danny, 149, 176
Keats, John, 59
"Keep Away from the Fellow Who
  Owns an Automobile," 53
Kentucky Club, 257
"Kentucky Sue," 87
Kern, Jerome, 3, 9, 33, 37–41, 46, 47,
  68, 87, 95, 105, 125, 130, 144, 149,
  150–51, 163, 183, 186–88, 196,
  199, 209, 215, 218–21, 229, 246,
  263, 264, 276, 277
Kiss Me, Kate, 178, 277
Knickerbocker Holiday, 200
Koehler, Ted, 3, 83, 245–50, 251,
  254, 256, 270, 276
Kurtz, Manny, 258–59

Lady, Be Good!, 16, 88, 103, 128–31,
  134
"Lady Be Good" (*see* "Oh, Lady Be
  Good!")
Lady in the Dark, 7, 149–50
"Lady Is a Tramp, The," 119, 122
"Lady of the Evening," 61
Lahr, Bert, 176
La-La Lucille, 128
Lane, Burton, 210–12, 250
Langer, Susanne, 4
Lardner, Ring, 165
"Last Time I Saw Paris, The," 188, 263
Laura, 277
"Laura," 277
"Lazybones," 265
Leave It to Me, 176
Lee, Linda, 157
Lee, Peggy, 252
Lehár, Franz, 31, 116
Leisen, Mitchell, 227
Lenox, Jean, 32
Let 'Em Eat Cake, 140
"Let's Be Buddies," 176
"Let's Call the Whole Thing Off,"
  144, 147
"Let's Do It," 16, 104, 153, 154, 160–
  63, 165
Let's Face It, 176, 222

"Let's Fall in Love," 249–50
"Let's Misbehave," 160–61, 162
"Let's Not Talk About Love," 176
Levy, Eugene D., 27n
Lewis, Sam, 84
Life Begins at 8:40, 196, 208, 250
"Life Is Just a Bowl of Cherries," 93
Life magazine, 7
Li'l Abner, 277
Little Egypt, 53
"Little Girl Blue," 117–18
Little Johnny Jones, 30
"Little Lost Child, The," 14
"Little More of Your Amor, A," 230
"Little Old Church in England, A," 67
Little Show, The, 195, 197
"Liza," 82
Lonely Romeo, A, 98
"Lonesomest Girl in Town, The," 216
"Long Ago and Far Away," 150
Longfellow, Henry Wadsworth, 61
"Look for the Silver Lining," 87, 92
Loos, Anita, 229
Losch, Tillie, 164
"Louise," 224
Louisiana Purchase, 67
"Love in Bloom," 226
"Love Is a Many-Splendored Thing,"
  260
"Love Is Here to Stay," 145, 148–49
"Love Is Just Around the Corner," 8–
  9, 225–26
"Love Me Forever," 82
"Love Me or Leave Me," 81–82, 83
Love Me Tonight, 114–16
"Love Walked In," 145
Lovelace, Richard, 6
"Loveliness of You, The," 241
"Lovely to Look At," 219
"Lover," 115
Lubitsch, Ernst, 225
"Lullaby of Broadway," 239
Luther, Martin, 13
"Lydia, the Tattooed Lady," 208
Lyrics on Several Occasions, 5, 13, 132,
  141, 146

MGM, 172, 203, 278
MacArthur, Charles, 207
McCarthy, Joseph, 33–34, 43–44, 112
MacDonald, Ballard, 88

MacDonald, Jeanette, 115, 225
McGinley, Phyllis, 8
McHugh, Jimmy, 215–19, 245, 250
Mackay, Ellin, 55
"Make Believe," 183–84
"Makin' Whoopee," 80–81
Mamoulian, Rouben, 115, 225
"Man I Love, The," 41, 110–11, 131,
  133
"Man That Got Away, The," 110–11,
  151–52
Mancini, Henry, 278
"Manhattan," 10, 16, 94, 99–103,
  104, 125, 163, 182
Manhattan Melodrama, 116
"Marie from Sunny Italy," 48
Marks, Edward, 20, 21
Marquis, Don, 8, 127
Martin, Mary, 176
Marx Brothers, 208
Marx, Samuel, 95n, 105n, 122n
"Massa's in de Cold, Cold Groun'," 20
Mast, Gerald, 9, 37, 56, 69, 130, 145n,
  177, 193, 220, 225, 234, 238, 280
Maxwell, Elsa, 155
Mayer, Louis B., 172–73, 180
"Mean to Me," 84–85
Meiser, Edith, 99
"Memories," 76
"Memphis Blues, The," 35
Mencken, H.L., 11, 109, 127
Mercer, Johnny, 3, 37, 78, 153, 211,
  250, 254, 262, 263–82
Merman, Ethel, 47n, 70, 93, 136, 170,
  176
Merry Widow, The, 116
Mexican Hayride, 177
Meyer, Joseph, 75
"Midnight Sun," 281
Midsummer Night's Dream, A, 178
Millay, Edna St. Vincent, 10, 242
Mills, Irving, 215, 255, 257–59, 261,
  262
Mills, Kerry, 27
"Mine," 140
"Mister Johnson, Turn Me Loose," 26
"Mon Homme," 112
Monaco, Jimmy, 33, 243
Monte Carlo, 225
"Mood Indigo," 259
"Moon River," 264, 278

Moore, Marianne, 10, 105
Morath, Max, 28
Mordden, Ethan, 94, 231, 234n, 235n, 238n
"More Than You Know," 85
Morley, Christopher, 8
Morris, James, 128–29, 252n, 275n
Morse, Lee, 112
"Most Beautiful Girl in the World, The," 118
"Mountain Greenery," 5, 104–5
Murphy, Gerald, 157, 159
Murphy, Sara, 157
Music Box, 54
"Music Makes Me," 82
"My Black Baby Mine," 26
"My Blue Heaven," 86–87
"My Funny Valentine," 119
"My Gal Is a High Born Lady," 27
"My Heart Belongs to Daddy," 176
"My Heart Stood Still," 96, 108–9, 118, 121
"My Ideal," 224
"My Man's Gone Now," 142
"My Mariuccia Take a Steamboat," 48
"My Mistress' Eyes Are Nothing Like the Sun" (Shakespeare sonnet 130), 90
"My Mother Would Love You," 176
"My Shining Hour," 274–75
"My Ship," 150

"Nagasaki," 237
Nash, Ogden, 8
National Winter Garden, 33
Nemo, Henry, 259
"Never Gonna Dance," 221–22
New Yorker magazine, 17, 124
Nice Going, 229
"Nice Work If You Can Get It," 144, 148
"Night and Day," 11, 155, 164–65, 171, 276
Noble, Ray, 251
"Nobody Knows the Trouble I've Seen," 27

Of Thee I Sing, 44, 67, 139–40
"Oh, Bright, Fair Dream" 298
"Oh! How I Hate to Get Up in the Morning," 47, 52, 67

"Oh, Lady Be Good!," 130–31
Oh, Look!, 43
"Oh Promise Me," 32
"Oh, That Beautiful Rag," 49
"Oh, What a Beautiful Mornin'," 190–91
O'Hara, John, 123, 176
Oklahoma!, 18, 41, 67, 68, 69, 117, 125, 140, 149, 177, 181, 186, 187, 189–94, 263, 275
"Old Devil Moon, 211–12
"Old-Fashioned Garden," 154, 158–59
"Old-Fashioned Tune Always Is New, An," 67
"Old-Fashioned Waltz, The," 159
"Ol' Man River," 3, 184–86, 191
"Olga, Come Back to the Volga," 159
"On and On," 11
"On the Atchison, Topeka, and the Santa Fe," 278
"On the Sunny Side of the Street," 217–18
On Your Toes, 117
"One for My Baby," 254, 273–74
"One I Love Belongs to Somebody Else, The," 79–80
One Night of Love, 83
"One Night of Love," 83
O'Neill, Eugene, 99
"Orchids in the Moonlight," 82
Our Town, 125
"Out of Breath," 264
Out of This World, 178
"Over the Rainbow," 3, 209–10, 274
"Over There," 52, 204
Oxford Anthology of Light Verse, 8

Pal Joey, 123–24
Palmer, Tony, 41n, 196n, 279n
Panama Hattie, 176
Paramount Pictures, 224, 229
"Pardon Came Too Late, The," 22
Pardon My English, 140
"Pardon My Southern Accent," 265
Paris, 16, 160
"Paris Is a Paradise for Coons," 33
Parish, Mitchell, 4, 255–58, 280
Parker, Dorothy, 8, 18, 197, 231, 234, 236
Pastor, Tony, 19, 22, 48
"Payola," 21

Pelham Cafe, 48
"People Will Say We're in Love," 191–92
Pessen, Edward, 15n
Philippe-Gerard, M., 280
"Picture That Is Turned to the Wall, The," 22
Pirate, The, 278
Platters, The, 187
"Play a Simple Melody," 50–51
Poetry magazine, 51, 255
Porgy and Bess, 140–42, 144, 275
Porter, Cole, 4, 7, 8, 9, 11, 15, 16, 17, 18, 42, 47, 63, 70, 71, 73, 90, 93, 94, 97, 103, 117, 120, 124, 138, 143, 153–80, 181, 188, 189, 192, 196, 198, 199, 200, 213, 215, 222, 227, 228, 231, 253, 264, 265, 266, 268, 269, 272, 276, 281
Potter, Paul M., 127–28
Pound, Ezra, 227
"Prayer," 116
Present Arms, 109–11
"Pretty Girl Is Like a Melody, A," 47, 53–54
Princess Theatre, 39–40, 95, 99
Pulitzer Prizes, 125, 140
Puttin' on the Ritz, 62
"Puttin' on the Ritz," 62–63, 65, 66, 168

RCA, 279
RKO, 144
Rafe's Paradise, 245
"Ragtime Jockey Man," 53
"Ragtime Soldier Man," 53
"Ragtime Violin," 53
Rainey, Ma, 36, 245
Rainger, Ralph, 8, 226–29
Raksin, David, 277
Razaf, Andy, 3, 87
"Real American Folk Song, The," 127–28
Red, Hot and Blue, 165
"Red Hot Coon, A," 29
"Red, Red Rose, A," 6
"Red Wheelbarrow, The," 11, 102
Redman, Don, 255
Redmond, Johnny, 259
"Remember," 55
Remick, Jerome, 21, 233, 246

Revel, Harry, 240
Revenge with Music, 196
"Rhapsody in Blue," 128
Riggs, Rolla Lynn, 125, 187, 189–90
Roberta, 187, 219
Robbins, Jack, 116
Robin, Leo, 2, 9, 213, 214, 223–30, 237
Rodgers, Richard, 5, 10, 16, 18, 41, 68, 69, 73, 94, 95–125, 140, 145, 149, 155, 160, 163, 165, 181, 182, 189–94, 206, 213, 215, 216, 234
Rogers, Ginger, 47, 82–83, 137, 144, 220–22
Rogers, Roy, 177
Romberg, Sigmund, 37, 73, 98, 182, 215
Ronell, Ann, 250–51
Roosevelt, Franklin Delano, 204, 210
"Rosalie," 173
Rosalie (Broadway musical), 131
Rosalie (Hollywood film), 172
Rose, Billy, 85, 117
Rose, Fred, 83
Rose Marie, 182
"Rose of the Rio Grande," 237
Rosenfeld, Monroe H., 19, 22, 79
Rouault, Georges, 129
Rourke, Michael E., 38–39, 61
Ruby, Harry, 236, 237, 240
Runyon, Damon, 217
Russell, Bob, 260–61, 262
Ryskind, Morrie, 139–40

" 'S Wonderful," 133–35, 150, 266
"Saga of Jenny, The," 149
"St. Louis Blues," 35–36, 271
St. Louis Woman, 275–77
Salter, "Nigger Mike," 48
Salvin's Plantation Club, 244
"San Francisco," 82
"San Francisco Bound," 53
"Santa Claus Is Coming to Town," 252
Saratoga, 277
"Satin Doll," 262, 281–82
"Say It Isn't So," 11, 47, 60–61
"Say It with Music," 11, 54
Scandals (George White's), 88, 89, 92
Schertzinger, Victor, 83
Scheurer, Timothy, 64n

Schubert, Franz, 180
Schwartz, Arthur, 94, 96, 98, 195–203
Schwartz, Charles, 154, 158
Scott, Randolph, 188
"Second Hand Rose," 233
See America First, 157–58
Segal, Vivienne, 122–23
"Sentimental Me," 103
"September in the Rain," 240
"September Song," 200
"Serenade in Blue," 243
Seven Brides for Seven Brothers, 278
Shakespeare, William, 14, 90, 109, 118, 178
"Shaking the Blues Away," 61
Shall We Dance?, 144, 145, 147, 148
Shapiro, Elliot, 98, 99
Shapiro, Maurice, 20
Shaw, Arnold, 37
Shelley, Percy Bysshe, 5
"She's Funny That Way," 37
"She's Getting Mo' Like the White Folks Every Day," 27
"She's My Warm Baby," 29
"Shine On, Harvest Moon," 32
Show Boat, 41, 140, 181, 183–87, 189
Show Girl, 82
Shubert Brothers, 39
"Side By Side," 86
Silk Stockings, 179
Silvers, Louis, 83n
Simple Simon, 111–14
Sinatra, Frank, 246, 254, 279
Singin' in the Rain, 235
Singing Fool, The, 92, 233
"Si's Been Drinking Cider," 53
Sky's the Limit, The, 274–75
Smart Set, 7, 127
Smith, Bessie, 36, 245
Smith, Kate, 47
"Smile and Show Your Dimple," 63
"Smoke Gets in Your Eyes," 187
"So in Love," 178
"Solitude," 258
"Some of These Days," 33
"Some Sunday Morning," 250
"Somebody Loves Me," 88
"Someone to Watch Over Me," 133
Something for the Boys, 177, 222
"Something to Remember You By," 199–200, 201

"Something's Gotta Give," 278
Sondheim, Stephen, 42, 182
"Song Is Ended, The," 63
"Song Is You, The," 188
"Sonny Boy," 92
"Soon," 139
"Sophisticated Lady," 257–58
South Pacific, 277
Spaeth, Sigmund, 23n, 27, 92
"Spring Is in My Heart Again," 265
"Star Dust," 4, 245, 255–57
Star Is Born, A, 151–52
"Stars Fell on Alabama," 257
Stein, Gertrude, 62
Steinbeck, John, 125, 189
Stern, Joseph, 20
Stewart, Lawrence D., 127n, 148
"Stop That Rag," 46
"Stormy Weather," 3, 151, 248–49, 276
Strike Up the Band, 131, 139
Student Prince, The, 215
Styne, Jule, 3, 229–30
Sullivan Arthur, 5, 6, 13, 30, 39–40, 139, 149, 156
Summer Holiday, 278
"Summer Wind," 280
"Summertime," 141
"Sunny Disposish," 134
Sunny River, 181
"Sunny Side Up," 92
"Supper Time," 63, 249
"Sure Thing," 11
"Surrey with the Fringe on Top, The," 192
Sutton, Harry, 32
"Swanee," 75, 128
Sweet Adeline, 187
Sweet Charity, 222–23, 229
"Sweet Leilani," 147
"Sweet Lorraine," 255
"Sweet Sue," 90
Swing Time, 220–22

Take a Chance, 93
"Take Me in Your Arms," 257
Tanguay, Eva, 32
Taylor, Deems, 159
"Tea for Two," 72–75, 90, 235
Temple, Shirley, 93, 209
"Temptation," 236

"Ten Cents a Dance," 111–14
Texas Li'l Darlin', 277
Thalberg, Irving, 213, 235
"Thanks a Million," 82
"Thanks for the Memory," 17, 227–29
"That Certain Feeling," 132
"That Hula-Hula," 53
"That International Rag," 46
"That Old Black Magic," 3, 78, 211, 272–73
"That Opera Rag," 49
"That's Entertainment," 203
Theater Guild, 99, 141
"There Will Never Be Another You," 242
"There'll Be Some Changes Made," 11
"There's No Business Like Show Business," 69
"They All Laughed," 12, 126–27, 144, 148
"They Can't Take That Away from Me," 11, 14, 144, 145–47, 199
"They Didn't Believe Me," 37–39, 61, 130, 264
"They Say It's Wonderful," 70
"Things Are Looking Up," 144, 147
"This Can't Be Love," 15, 121
This Is the Army, 67 177
"This Is the Army, Mr. Jones," 47
"This Is the Life," 53
Thomas, Danny, 14
"Thou Swell," 70, 106–8
"Three Little Words," 236
Three Sisters, 219
"Till the Clouds Roll By," 40
"Tip Toe Through the Tulips," 237–38
Tip-Toes, 132
"To My Mammy," 60
Too Many Girls, 122
"Too Marvelous for Words," 266–67
Top Banana, 277
Top Hat, 65
"Top Hat, White Tie and Tails," 47, 65–66
Tree Grows in Brooklyn, A, 203, 222
Trip to Chinatown, A, 41
"True Blue Lou," 224–25
"True Love," 154, 179–80
"Tschaikowsky," 7, 149

Tucker, Sophie, 33
Turk, Roy, 84
Twain, Mark, 5, 106–7, 109
"'Twas Only an Irishman's Dream," 237
Twelfth Night, 14
Twentieth Century-Fox, 229, 240, 243
"Twinkle in Your Eye, A," 122
"Two Little Babes in the Wood," 159

"Under the Bamboo Tree," 27–28, 62
Untermeyer, Louis, 8
Updike, John, 8

Vallee, Rudy, 47, 60
Van Alstyne, Egbert, 76
Vanderbilt Revue, 217
Vanity Fair, 7
Variety, 103, 181
Vers de Société Anthology, 6–7
"Very Precious Love, A," 260
"Very Thought of You, The," 251
"Vilia," 116
Von Tilzer, Albert, 87

Wake Up and Dream, 164
Walk a Little Faster, 205
Waller, Thomas ("Fats"), 89
Wallichs, Glen, 279
"Warmest Baby in the Bunch, The," 30
"Warmest Colored Gal in Town, The," 29–30
Warner Brothers, 93, 233, 238, 240
Warren, Harry, 231, 237–43, 246, 266, 267
Waste Land, The, 58
Waters, Ethel, 63, 244–45, 248
Watch Your Step, 50
"Way You Look Tonight, The," 221
"We Only Pass This Way One Time," 211
Webb, Clifton, 197–98
Webster, Paul Francis, 260, 261, 262
Weill, Kurt, 7, 149–50, 200
Wells, Carolyn, 6–7
"We're a Bunch of Nonentities," 156
"We're in the Money," 276
"What Are You Doing the Rest of Your Life?," 250

"What Can You Say in a Love Song?," 15, 208

"What Is This Thing Called Love?," 164

What Price Glory?, 233

"What'll I Do?," 56–58, 77–78

"What's New?," 253

"When I Lost You," 51, 55

"When I Used to Lead the Ballet," 156

"When I'm Not Near the Girl I Love," 210

"When It's Night Time in Dixie Land," 53

"When My Sugar Walks Down the Street," 216

"When the Girl You Love Is Many Miles Away," 29

"When the Idle Poor Become the Idle Rich," 210

"When the Midnight Choo-Choo Leaves for Alabam'," 53

"When the World Was Young," 280–81

"When You Want 'Em, You Can't Get 'Em, When You've Got 'Em, You Don't Want 'Em," 127

"When You're Down in Louisville," 53

"Where Is the Life That Late I Led?," 176

"Where Is the Song of Songs for Me?," 63

"Where or When," 119–20

"Where's That Rainbow?," 106, 216

Whistler, James McNeill, 128, 166

"Whistling Rag," 53

White, E.B., 8

White, George, 86, 89

"White Christmas," 47, 68

Whitman, Paul, 233, 250, 264, 268

Whiting, George, 87

Whiting, Richard, 76, 224–25, 266

Whoopee, 80, 81

"Why Do I Love You?," 183, 184

Wilder, Alec, 33, 35, 53, 60, 64, 72, 73, 78, 87, 96n, 106, 112, 120, 161, 163, 171, 172, 199, 200, 212, 218, 237, 240, 242, 248, 252, 253, 255, 257, 259, 260, 268, 272

Wiley, Lee, 136

Wilhelm, James, 15

Williams, William Carlos, 11, 102

"Willow Weep for Me," 250–51

Winchell, Walter, 80, 214

Winkle Town, 99

"Witchcraft," 211

"With a Song in My Heart," 111

"With All Her Faults I Love Her Still," 79

Within the Quota, 159

Witmark Brothers, 20, 233

Wizard of Oz, The, 209–10

Wodehouse, P.G., 39–40, 95, 100, 105, 184

Woods, Harry, 86

Woollcott, Alexander, 99

Woolley, Monty, 177

Wordsworth, William, 88

"Would You?," 235

"Would'ja for a Big Red Apple?," 265

Wynn, Ed, 112

"Yes, Sir! That's My Baby," 76, 77

"Yesterdays," 187

"Yiddle on Your Fiddle, Play Some Ragtime," 49

Yip, Yip Yaphank, 52, 67

"You and the Night and the Music," 201–2

"You Are Love," 184

"You Are My Lucky Star," 235

"You Are So Fair," 121

"You Are Too Beautiful," 116

"You Can't Get a Man with a Gun," 68, 69

"You Do Something to Me," 163–64, 272

"You Go to My Head," 17, 252–53

"You Made Me Love You," 33–34, 112

"You Must Have Been a Beautiful Baby," 267

You Never Know, 176

"You Stepped Out of a Dream," 82

"You Took Advantage of Me," 11, 110–11

"You Were Meant for Me," 235

"You'll Never Know," 242

Youmans, Vincent, 72–75, 82, 128, 224, 246

Young, Joe, 84, 244

"You're a Grand Old Flag," 31

"You're an Old Smoothie," 93
"You're Driving Me Crazy," 11
"You're Getting To Be a Habit with
  Me," 238
"You're Growing Cold, Cold, Cold," 30
"You're Just in Love," 70–71
"You're the Cream in My Coffee," 10,
  89–91
"You're the Top," 9, 89, 154, 156, 157,
  165–67, 168

"You've Been a Good Old Wagon But
  You Done Broke Down," 26

Zanuck, Darryl F., 238
Ziegfeld, Florenz, 32, 39, 47, 53, 54,
  80, 99, 112, 114, 142, 158, 183
Ziegfeld Follies of 1919, 53
Ziegfeld Follies of 1936, 142
Zukor, Adolph, 225